The App & Mobile Case Study Book

Ed. Rob Ford/Julius Wiedemann

TASCHEN

"Logic will get you from A to B.
Imagination will take you everywhere."

Contents

Icon Key

AIR   AIR App

An   Android App

Bl   BlackBerry App

Fa   Facebook

In   Intel App

iP+   iPad App

iP   iPhone App

m.   Mobile Website

Ovi   Ovi App

Pa   Palm App

Sy   Symbian App

Vi   Video Demonstration

w.   Website

W7   Windows 7 App

Project Web Link

www.taschen.com/tamcsb

Important note regarding the icons and links in this book:
As readers will be aware, many apps and mobile projects are available across multiple devices and platforms. Rather than confuse each case study with multiple links, we have created a single link www.taschen.com/tamcsb which readers can access to see all available devices, video demos and more for each app and mobile project in this book.

We have also created a simple list of icons so you can quickly identify what devices each app or mobile project is available on and whether or not there are additional items of content available to view, i.e. video demonstrations, etc.

Please familiarise yourself with the icon key and also the direct link for all cases and associated content in this book.

Foreword

by Ralph Simon,
Mobilium International

Foreword Foreword

Never before in human history has a medium and technology grown as fast, become as widely used and globally widespread as the mobile phone and the growing phenomenon of mobile device technologies. In less than two decades, the number of mobile phones and users has grown to nearly 5,000,000,000 active mobile users and handsets around the world.

Mobile devices have also ushered in a whole new pattern of social interaction, mobile lifestyles and uses that would have been totally unthinkable to Alexander Graham Bell, the eminent scientist, inventor and engineer, who is credited with inventing the first practical telephone and was awarded the first US patent for the telephone in 1876. Would he have ever imagined that by the start of the 21st century, every facet of modern life and its conduct would be touched by the impact of the ubiquitous mobile device and its social impact?

The evolution of mobile communications and wireless enablement has spread way beyond mere voice calls. The modern age has seen a mushrooming of usage, applications and smart computer chips that allow a profound versatility for the technologies that drive it.

"In the formative years of mobile usage and adoption in the 1980s, Japan and Korea led the way with versatile telecom networks and services that soon caught the public imagination."

In the formative years of mobile usage and adoption in the 1980s, Japan and Korea led the way with versatile telecom networks and services that soon caught the public imagination, inculcating a swift reliance on their capabilities and becoming a compulsory tool for urban dwellers.

Europe soon followed and before too long, rapidly developing usage and networks saw a huge public demand and use of text messaging, entertainment applications and colour screens. North America was not immune from the swift adoption of mobile phone technologies: Canada and then the USA followed the pattern of fast growth and development that had been seen in Asia and Europe.

"Mobile required a totally different design and approach from the Internet and a fundamental shift occurred in January 2007, when smartphones and the iPhone were introduced."

By 2000, the die had been cast. People wanted to be connected, as text and multi-media messaging grew exponentially. Nowhere was this more dramatic and evident than in the emerging markets, economies and countries. Africa, South America and the rest of Asia saw massive adoption of mobile phone usage, leapfrogging old-style wire-line telecommunication for fully wireless implementation and operation. Subscriber numbers grew with massive speed and political impact. A whole new industry of applications developers, platform architects, device design and ingenious coding acted as a further lightning rod for the mobile industry's rise and rise.

Economists could accurately project a country's Gross Domestic Product growth by the increase in new subscribers. From these accelerated beginnings, massive new telco industries grew with business and social usage of mobile phones producing billions in revenue for all in the mobile value chain.

Mobile required a totally different design and approach from the Internet and a fundamental shift occurred in January 2007, when smartphones and the iPhone were introduced, giving high-powered computer capabilities to the mobile phone user. This was a critical junction point in the evolution of mobile usage, which further influenced the growth and popularity of applications specially made for smartphone mobile devices.

"The dizzying pace of innovation and an always improving user experience guarantee that the next decade will see every facet of modern living touched by mobile."

The future promises continued fast-paced development of personalised applications and the potential for changing the face of health-care, mobile banking, rural and distance education, entertainment, mobile games and enabling mobile lifestyles and social interaction as never seen before.

The dizzying pace of innovation and an always improving user experience guarantee that the next decade will see every facet of modern living touched by mobile as applications and mobile machine-to-machine connectivity becomes the dynamo driving modern life.

Ralph Simon
Mobilium International
Emeritus Mobile Entertainment Forum – Americas

Bio.
<u>Ralph Simon</u>
CEO, Mobilium International
Emeritus Mobile
Entertainment
Forum – Americas

Ralph Simon predicted in 1997 that mobile phones would become the indispensable voice/social networking companion for consumers and their increasingly mobile lifestyles around the world. He became known as the founder of the global mobile entertainment industry, introducing ring tones into the Americas, Europe, UK, Australasia and Africa, and became known as "the Father of the Ring Tone".

Today, he advises companies, brands, governments and individuals worldwide seeking to innovate, using mobile as a platform for entertainment, health, education and mobile payments. He has received wide international recognition for his expertise and far-sighted contributions to the mobile industry since 2005.

Based in London, he is a Fellow of the Royal Society of Arts in the UK and a member of the National Academy of Recorded Arts & Sciences in the USA.
–
www.mobilium.com
www.m-e-f.org

"Never before has a medium and technology grown as fast and become as globally widespread as the mobile phone."

Introduction

by Rob Ford,
Favourite Website Awards (FWA)

Here we are with the third in this, now bestselling, series following on from *Guidelines for Online Success* and *The Internet Case Study Book*. This book follows on closely and in a very similar style and layout to *The Internet Case Study Book*. One extra you will find in this book is a link with most of the case studies for a video demo of the featured mobile project. In such a fast-changing world, I am acutely aware of the fact that mobile content is changing so fast and I was determined to make this book readable even in ten years' time. So, as long as YouTube and Vimeo stay with us, you should be able to watch a video demo of the cases showcased in this book, even if the project has changed or is no longer live.

"Something I have realised now I am on my third full book with TASCHEN is how they will become historical references in years to come."

Something I have realised now I am on my third full book with TASCHEN is how they will become historical references in years to come. Some people say that print is dead… I even have myself but I have also realised that certain types of books are now more valuable than ever, from a resource point of view. Many, especially the high-quality and beautifully crafted ones, are seen by many as objects now, rather than pure printed matter. As the publishing world evolves, we might see pure text-books die whilst books such as the ones in this series become important for archiving eras past, especially in the fast-moving interactive world. In just ten years the web has evolved out of sight and now mobile has really taken over; ten years from now we will pick up this book and look with a wry smile at what we thought was mind-blowing at the time.

One of the biggest hurdles I faced when compiling this book was putting together a wide range of content that would appeal to any audience. I wanted to make sure I had a selection of some of the best mobile projects ever, with a good number of new and smaller projects, some by individuals with no budget apart from their own time and talent. Some cases have been extremely difficult to get to the book as they took a lot of pleading on my part to get them here as a case study. There were certain "must have" cases that I had to pursue until we had them here, and I'm pleased to say that I achieved everything I set out to with this book.

As mobile is so heavily influenced by mobile game apps right now, it makes sense to start the book with a chapter on "Games". We have Joe Wee of Chillingo, who has published a wide range of successful titles on the App Store including Angry Birds and Cut the Rope, introducing the chapter and giving us an insight into where mobile gaming has been, where it is now and where it might be going.

Chapter Two is all about "m-commerce" and is introduced by Ian Wharton of Zolmo. Zolmo made the record-breaking Jamie Oliver's 20 Minute Meals app, which the chapter leads with. This chapter covers apps, mobile websites, iAds and an iPad advert. One of the apps in this chapter, The Elements: A Visual Exploration, shows the true power of mobile as its net revenue in the first six months was $1 million.

"As the publishing world evolves, we might see pure text-books die whilst books such as the ones in this series become important for archiving eras past, especially in the fast-moving interactive world."

Chapter Three focuses on "Promotional" mobile projects and is introduced by AKQA's Chairman, Ajaz Ahmed. This chapter kicks off with the Nike True City app, which was the first iPhone app ever to be awarded an FWA Site of the Day, which Nike launched in support of the re-release of the classic and updated AM1 shoes (in 1989, Nike designer Tinker Hatfield created the original Air Max 1). This chapter covers everything from the official The Rules of Golf app, Stephen Fry's own app and the only app ever to be seen on Oprah Winfrey (Alice for the iPad).

"Social" is Chapter Four and gets introduced by the Founder and CEO of TweetDeck, Iain Dodsworth. We also have a case study on the Android version of TweetDeck. TweetDeck is my most used app for sure and Iain really does give us a great intro and some thought-provoking words in this whole new world we call "social". You'll find another eclectic mix with some famous case studies for Flipboard and Foursquare, plus one that really blew me away when I first saw it, The World Park, which used QR codes and image scanning to create awareness and engagement for New York's Central Park. My dog also makes a cameo appearance in this chapter in the images for the LEGO Photo app!

The final chapter is "Utilities" and this is introduced by Remon Tijssen from Adobe. Utilities could also be called "Miscellaneous" as this chapter heaps in a very wide variety of mobile projects. In fact Remon does talk about what utilities always were recognised as and what they are now being considered to be.

As with all the chapters in the book, this one begins with a case study connected to the person who wrote the chapter introduction, so this one leads off with Adobe Ideas. The Adobe Ideas case study includes a great story of how an individual, Sylvester Cann IV, was at a rally at the University of Washington, in Seattle, USA, and managed to get President Obama to autograph his iPad.

"One thing that is certain about mobile, it is taking over many of our lives. That's not as a telephone but as a way of keeping in the loop and managing our daily routines."

If you are familiar with this series of books you will know that Lars Bastholm from Ogilvy has written the Afterword for both previous books. His summing-up and great personality always rounds each book off so perfectly and I was lucky enough to get Lars to sum this book up as well. Lars always provides his unique way of writing and I do love to quote him…

"I got my first mobile phone in 1995. It was huge. It made phone calls, and that was it… If you put it in your jacket pocket, it looked like you were packing heat." – Lars Bastholm, 2010.

I hope you'll enjoy this book as much as the previous ones if you have both or either of them. If it is your first I hope the way we have laid things out makes it an easy to browse, yet informative book. Each case study gives you the credits, awards won and then the case itself: The Brief; The Challenge; The Solution; The Results. Plus, there are some Hot Stats so you can see what results were achieved easily and quickly. You'll also find a quote on every case, with some from the likes of Steve Jobs and Stephen Fry. Please make sure you see page 8 which shows you a key to the icons used in this book and also shows the generic link for all the mobile project links in this book.

One thing that is certain about mobile, it is taking over many of our lives. That's not as a telephone but as a way of keeping in the loop and managing our daily routines. As @tadd31 said on Twitter, "where is the app to order the turkey?"

So, without further ado and in the words of myself… do read this book, don't let it collect dust!

Rob Ford
The FWA Network

Bio.
Rob Ford
Founder & Principal,
The FWA Network

Rob Ford, born England 1969, founded Favourite Website Awards (FWA) in May 2000, a recognition program for cutting-edge web design which has since served over 100 million site visits.

His work has been featured in numerous publications including *The Chicago Tribune*, *The Guardian*, *Penthouse*, *The Big Issue* and many web-related magazines. He has judged for most of the industry award shows, contributes regularly to other well-known webdesign sites and magazines (from all corners of the globe) and has been writing a regular column for Adobe since 2006.

The FWA network showcases not only cutting-edge websites but also the best in mobile via the FWA Mobile Showcase and the best in photography via FWA Photo.
–
www.thefwa.net
www.robford.com
Twitter @fwa

Do
read this book.
Don't
let it collect dust!

Games

Introduction by
<u>Joe Wee</u>, Chillingo Ltd

01

The mobile gaming industry has experienced several seismic levels of changes in the last decade alone. The driving forces include technological advancements of the handsets, the ubiquity of mobile broadband, dramatic changes of the mobile value chain itself, shifts in consumer demand and expectations and last but not least the evolutions of the games themselves.

Mobile handsets are becoming more feature rich with enhanced capabilities. Higher-end devices are more conducive to gaming. The mobile phone has transformed from being a mere telecommunications tool, to a multi-purpose gadget that is with you any time and anywhere – that you can use for your gaming needs. Its impact on the handheld gaming market is significant. Morgan Stanley, August 2010, reports that the iPhone has already become the second-largest handheld gaming platform in terms of installed base and is set to surpass global leader Nintendo by the end of 2011.

"In 2010, 64 million people will play mobile games (at least monthly), a number that will rise to 94.9 million by 2014."

43.8 percent of the portable gaming market now uses a phone to play games on – a rise of more than 53 percent over the course of the year according to Interpret Analyst Firm, November 2010. The addressable market for mobile games will increase dramatically. User adoption towards higher-end devices is happening rapidly, and we see that trend continuing. According to eMarketer, August 2010, casual gaming on the mobile platform has driven adoption of mobile games to more than a quarter of mobile subscribers and more than one in five members of the US population. In 2010, 64 million people will play mobile games (at least monthly), a number that will rise to 94.9 million by 2014.

Mobile broadband has come a long way since the old GSM days. Its contribution to mobile gaming cannot be overestimated. It has been the main driving force that enabled digital distribution of mobile games. The advent of Wifi-based technology and 3G had allowed for immediate and "cheap" over-the-air downloading of content. If a game is under 20Mb in size today, you would be able to download the game directly on to your iPhone from the App Store – this is truly phenomenal if you compare to only a few years ago where 1 to 2Mb would have been a significant fit.

Looking forward in terms of infrastructure, I foresee faster data speeds and all-you-can-eat mobile broadband deals. Carriers and handset manufacturers will be investing in ubiquitous billing infrastructures which are fully integrated into App Stores across all carriers and devices. Consumers are already trained to purchase and consume content "immediately".

The value chain for digital mobile games has evolved. Once upon a time, carriers controlled the mobile gaming market through their so-called walled garden ecosystems. Content was delivered, stocked, marketed and distributed directly to phone subscribers. It was a great business indeed. The advent of the first smartphone-based content saw the birth of direct consumer portals run independently of operators that tried to compete with carriers decks but with limited success.

"This all changed when Apple entered the market with the App Store, fully armed with over 100 million ready-to-go consumers who had already been trained how to purchase digital music via iTunes."

This all changed when Apple entered the market with the App Store, fully armed with over 100 million ready-to-go consumers who had already been trained for years in how to purchase digital music via iTunes. Overnight, the marketplace for mobile games was to change forever, completely revolutionising how consumers purchase and consume content on the go.

Today, Apple's App Store houses over 300,000 apps, and is home to some of the cleverest, most innovative and downright fun games anywhere in the industry. The end-user charging models are also evolving. From one-off billing to a multitude of monetization models purely driven by the growing consumer demand for free games. There will be new options for selling mobile games and engaging with consumers beyond the operator decks. The smartphone and tablet market will continue to grow, and there will almost certainly be other major "App Stores" that simplify the consumer purchasing process leading to further market growth.

The games themselves have evolved from a few well-placed pixels to visually stunning environments with complex and realistic physics. It is no longer the case that games on mobile format should have the caveat "but it's only a mobile game." Games on those original green and black-screened cell phones seem like ancient history. Mobile gaming is no longer seen as a nerdy, geeky form of entertainment. Now, the mobile gaming community is quite possibly the most varied demographic in all of video-game history. Everyone from three-year-olds to ninety-three-year-olds is playing mobile games, which has led to a multi-billion dollar industry.

Consumer expectation has changed; mobile phone gamers now want the content they pay for to meet a certain level of quality, and rightly so. Every time a good game is released the bar is raised slightly higher. The App Store in particular has a customer base that will not tolerate poor quality simply because their mobile phone isn't considered a "true gaming platform". The desire for polished, innovative and original content puts pressure on developers and publishers, of course, but it's certainly a positive thing; it forces the industry forward and we have to, and do, move along with it otherwise we risk being left behind.

"With new technology and ideas that can change the face of gaming around every corner, there are some very exciting times ahead for all of us."

The mobile gaming market has definitely reached a tipping point in its evolution, with tremendous innovation evolving and driving it forward. Something that will remain constant is the expectation of high-quality, original and fun games. Developers and publishers alike will not be allowed to take their foot off the gas for a second. It's one of the reasons passionate individuals such as myself remain in the industry, we never want to reach a point where we believe we can sit back and say, "That ought to do it." It's an absolute joy discovering new games and ideas; some of the most talented geniuses in the world are mobile game developers and the type of innovation they create is simply mind-blowing.

Nothing is set in stone of course, but with new technology and ideas that can change the face of gaming around every corner, there are some very exciting times ahead for all of us.

Joe Wee
Chillingo Ltd

Bio.
<u>Joe Wee</u>
Chillingo Ltd

Joe Wee is the Co-General
Manager of Chillingo, the
leading independent games
publisher and division of
Electronic Arts. Chillingo won
the Mobile Entertainment
award for Best Games
Company of 2010 and has
published a wide range of
successful titles on the App
Store including Angry Birds,
Cut the Rope, Predators
and Minigore.
–
www.chillingo.com

"The mobile gaming industry has experienced several seismic levels of changes in the last decade alone."

Cut the Rope

"We'd put it right up there with the best casual games on the App Store."
Chris Reed, Slide to Play

The Brief

Cut the Rope is one of the most innovative games on the App Store. Using cutting-edge physics-based gameplay, the object is to get the candy from the rope to the mouth of Om Nom, the cute little green creature with a serious sweet tooth.

The App Store is one of the fastest-growing markets in entertainment, and as such, consumers have come to expect a high level of quality and innovation in the games they purchase and play. When bringing out a new app now, mediocre simply doesn't cut it. It is incredibly important that the app is innovative, polished and above all, fun.

The Challenge

When the Chillingo Team first saw Cut the Rope, we instantly realized we had something special. Our challenge, as the publisher, was ensuring that the game was as polished as possible so that consumers would find Om Nom and his world as adorably addictive and cute as we did. We collaborated closely with ZeptoLab, the innovative indie developers behind Cut the Rope, and fine-tuned every detail prior to launch.

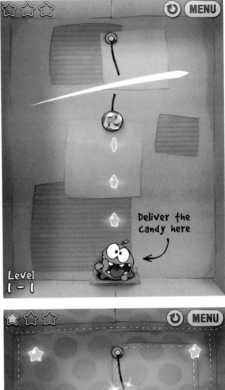

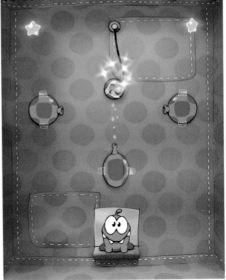

Developer
ZeptoLab

Publisher
Chillingo Ltd

Credits
Chillingo Ltd
www.chillingo.com
ZeptoLab
www.zeptolab.com

Awards
FWA Mobile

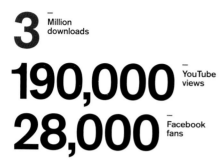

3 Million downloads

190,000 YouTube views

28,000 Facebook fans

The Solution

Chillingo performed rigorous Quality Assurance checks on Cut the Rope throughout several weeks and provided ZeptoLab with detailed feedback and advice to improve the gameplay mechanics and levels. ZeptoLab reciprocated by producing high-quality builds, and accepted our suggestions and advice to enhance the game.

Chillingo's PR and Marketing Team also meticulously implemented a launch schedule that would drive consumer and press interest to spike sales at launch. Not an easy feat given that the App Store has several thousand competitive apps. The team continued with social media efforts to engage consumers and invite them to download and play the game.

The Results

Cut the Rope was released on the App Store in October 2010 and was an instant success, reaching number one within just a few days. It subsequently broke the App Store record as the quickest app to reach one million sales, and then reached two million sales in just three weeks. It has since become one of the most popular apps on the store and a huge favorite among the press and consumers alike.

Since its release, Cut the Rope was inducted into the Apple App Store hall of fame and remains one of the highest-rated apps available today.

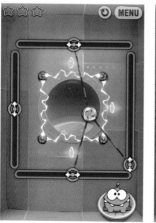

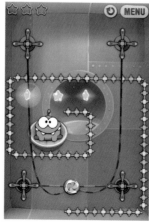

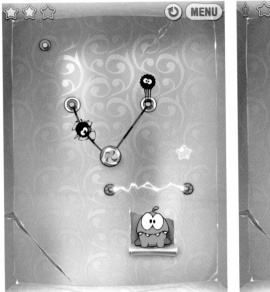

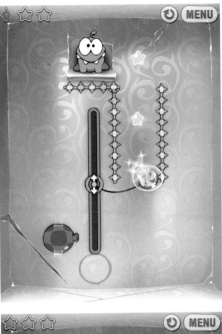

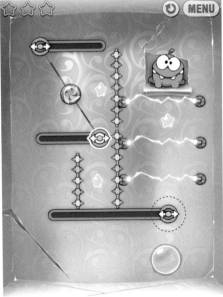

Angry Birds

"I think it's time to cut out the middle man and just have Angry Birds installed on every device. This is the most playable and addictive game of all time and everyone should experience it."
Rob Ford, FWA

The Brief

Rovio has a strong background in creating mobile games, with over 50 titles produced since 2003. With the turn of 2009, the smartphone market was emerging stronger than ever, and all reports indicated that portable and mobile gaming was growing significantly. The company decided that this was the perfect opportunity to create a completely new, original intellectual property, and a polished game around it. The iTunes App Store was the most mature established mobile marketplace at the time, and the devices had all the properties we wanted to develop for, so iPhone was a natural choice for the target device. Rovio CEO Mikael Hed and marketing manager Matthew Wilson laid out the desired parameters for the upcoming concept from Senior Game Designer Jaakko Iisalo.

The Challenge

In early 2009, designer Jaakko Iisalo was pitching a number of concepts to the team at the Rovio studio. One of the concepts was titled *Birds*, a somewhat complicated design which, however, included some very grumpy-looking bird characters. The studio loved the characters, and the decision was made to take the characters, and polish the game concept into something more accessible and universally appealing. Some of the key elements of the design include the use of intuitive touchscreen controls, physics-based gameplay that people can relate to, and a flexible scoring system, which caters to all players from complete novices to hardcore gamers.

Client
Rovio Ltd

Credits
Rovio Ltd
www.rovio.com

Awards
FWA Mobile; #1 Paid App in 71 countries; App Store Hall of Fame; 2010 Mashable Awards; CNET 100 top iPhone winner; Develop Award; Yahoo!

The Solution
The game concept was developed and polished over a number of iterations during eight months. The different birds and their special abilities were carefully balanced with other gameplay elements and level design, to create the perfect experience. Skill, luck, physics and destruction combine to create one of the most satisfying casual puzzle games ever made. To get the finished game noticed, Rovio went the creative way, and did simple things a bit differently from the competition. For example, no other successful apps in the App Store had a proper trailer, only gameplay videos. The studio produced a cinematic trailer for Angry Birds, and to this day the trailer has over 17 million views on YouTube.

The Results
In mid-February 2009, Angry Birds was featured in the UK App Store, and went Number One shortly thereafter. The game went on to dominate 71 App Stores around the world, with versions released for various platforms such as Palm WebOS, Android and Symbian^3. By December 2010, different versions of Angry Birds had been downloaded over 50 million times. In early 2011, the game was released for PC and Mac, and the developer has plans to create more Angry Birds games for every feasible platform. To sum it all up, many things clicked into place in the process of creating Angry Birds and making it the benchmark casual game for people of all ages. Rovio set out to create a hit game, and certainly did something right along the way.

75 Million downloads

20 Million iOS downloads

20 Million Android downloads

376 Thousand Facebook fans

3 Trillion Pigs popped in 2010

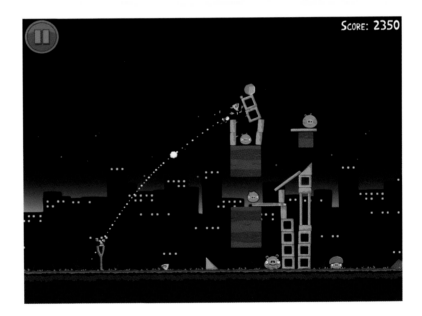

Fruit Ninja

"If you play it for ten seconds, then you'll know everything there is to know about Fruit Ninja. You'll know it's fun, and you'll know you want to keep playing."
Andrew Podolsky, Slide to Play

Client
Halfbrick

Credits
Halfbrick
www.halfbrick.com

Awards
FWA Mobile; PocketGamer Silver Award; IGN Most Addictive Game 2010; Nominated for IGN Mobile Game of the Year

The Brief

It couldn't be simpler. "Create a single-screen iPhone game." As the success of the iPhone and App Store continued to skyrocket, our designers at Halfbrick wanted to compete with the best of the best. It wasn't acceptable to flounder and struggle at the bottom of the charts – there were some out there making millions and serving the loyal iPhone community with truly iconic titles. As a developer focused on simple, casual downloadable titles, there was plenty of talent at Halfbrick ready to take on the App Store challenge. Being in Australia didn't matter. This was a global opportunity. We had to win.

The Challenge

What makes an iPhone game a success? Simplicity, addictiveness, personality, charm, mass appeal – just a few of many variables needing to be fine-tuned in order to make a big splash and get some serious downloads. There was also the challenge of marketing and PR, something far too many developers neglect – but we also knew that luck played a factor. Some amazing games never take off in the way that they deserve.

A genuinely iconic iPhone game doesn't come around that often, but it is possible to achieve with dedicated planning. In early 2010, this was the holy grail of game development, and not just in the mobile sector. We had the challenge of establishing ourselves among 200,000 others.

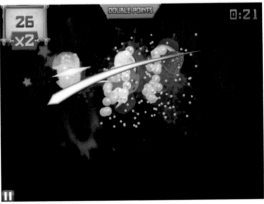

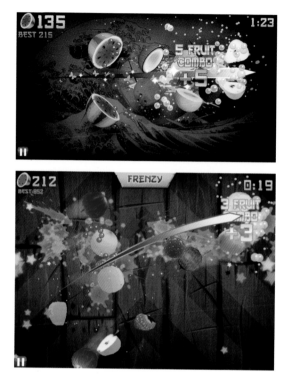

The Solution

Ninjas hate fruit! Three words to define a game. Luke Muscat, one of the lead designers at Halfbrick, pitched Fruit Ninja in a series of simple images, displaying watermelons and strawberries being sliced as they were launched up on screen. It had a universal theme that everyone could relate to, a one-touch method of learning and playing, and a brutally satisfying audio-visual experience, where players were rewarded for their efforts with a visceral set of juicy graphics and sounds.

The aim? Slice fruit to get a high score and avoid bombs which end your game. Simple, timeless satisfaction based on racking up scores and beating friends. For the iPhone market, and for addressing our initial brief – what could be better?

The Results

After an extensive marketing and PR campaign critical to launch, Fruit Ninja was an overnight success. It shot up the charts due to several factors – great reviews, great buzz and a well-timed App Store promotion. However, games come and go in the top ten all the time, but Fruit Ninja's core appeal and viral personality made it stick. It achieved the top-selling spot in multiple countries around the world, and although fluctuating from time to time, hasn't strayed from the top ten in most major markets since release.

By the end of 2010, including iPhone, iPad and Android sales, Fruit Ninja has been downloaded over four million times. We achieved exactly what we set out to achieve, and Halfbrick became an extremely successful and respected independent developer.

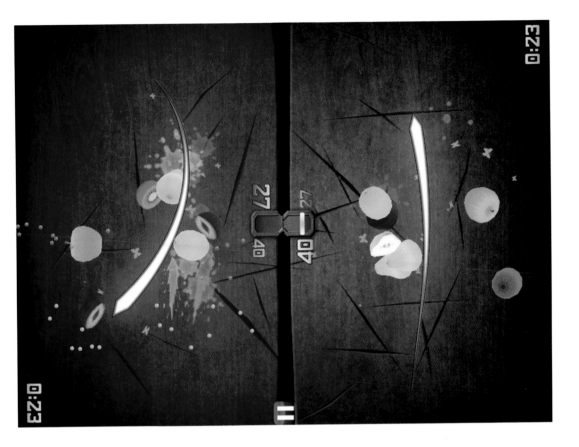

30 — Billion fruit sliced worldwide

74 — Days to achieve first million downloads

196 — Average player's high score

10 — Free updates since release

Plants vs. Zombies ^(aka PvZ)

Plants vs. Zombies ^(aka PvZ)

"It's just the sort of hyper-addictive game that you'd expect from indie publisher PopCap... And it's so crazy that it could only have come from a small developer with a strong creative vision untainted by big corporate strategy."
Wired

Client
PopCap Games

Credits
PopCap Games
www.popcap.com

Awards
25 Game of the Year nominations;
Editor's Choice Gold Award,
GameInformer (2010);
Editor's Choice Award, IGN.com
(2009); Editor's Choice Award,
GamePro (July 2009); Editor's Choice
Award, PC Gamer (June 2009);
Best Strategy Game of the Year,
USAToday.com (2009);
Casual Game of The Year (2009),
Chicago Sun-Times

The Brief

Originally designed and released by PopCap Games for both PC and Mac platforms, Plants vs. Zombies (PvZ) is a tower defense-style video game which pits a homeowner equipped with a wide range of plants against marauding zombies. With the extra help of neighbor "Crazy Dave", the player defends the house from attack from the front and back yards (the latter featuring a swimming-pool), from the roof, in daytime or at night.

PvZ has been the most successful new entrant in the PopCap product line since the release of Peggle in February 2007. An adaptation for the iPhone operating system was released in February 2010, and an HD adaptation launched for the iPad. An extended Xbox Live Arcade adaptation introducing new gameplay modes and features was released on September 8, 2010. PopCap has also announced a Nintendo DS iteration, released in February 2011, which will feature content unique to that platform. Furthermore, both the original Windows and Mac versions of the game have been re-released with additional content in a Game of the Year edition.

The Challenge

The budget for PvZ was modest compared to established PopCap games titles, and certainly compared to established industry titles. The achievement of word-of-mouth evangelism was the platinum standard for marketing success. The launch budget for the online version of PvZ was $60K and was targeted at the hardcore and consumer press ($20K). Another $35K was spent to build interest with the PopCap.com installed base. An additional $5K was allocated to create a great game video that was broadly circulated on portals and within social networks.

The budget for second half 2009 marketing of PvZ was not allocated save for a modest $15K of additional PR in the October timeframe. No retail variable marketing budget was allocated either. Additional budget and the probability of a fall retail launch was based on the initial success of the game, product reviews, customer reception, third-party interest and PopCap's collective abilities to virally spread the game in the late spring and early summer.

At a high level on the production side, some of the areas where significant time was spent getting things right in the iPhone/iPad adaptations of the PC/Mac versions were:
• Controls (touch interface vs. mouse multitouch on iPad)
• UI & Screen Layout (new seed packets, updated orientation, etc.)
• Game Performance (time was spent optimizing so the game would run at acceptable speeds on the slower devices)
• Art (the zombies' heads/faces were enlarged on the phone because they're a really iconic part of the game, and were too small on the phone when scaled down as much as the bodies).

The Solution

Gain attention and coverage from consumer and gamer press and analysts/influencers:
- Clever and informative press kits
- Enable early game trailer that shows depth and humor of game.

Evangelize long-lead press by providing a preview version of the game:
- Six long-lead publications provided in-depth reviews that hit the street as early as the first week of April.

Gain trial with core PopCap faithful:
- Heavy promotion on PopCap & Steam
- Use of PopCap newsletters and viral mechanisms.

Gain trial with secondary (teen) audience:
- Promotional ads on key target sites.

Ensure that all who try the game know/ understand depth of game beyond the first 15-minute experience:
- In game preview trailer
- Showcasing all the plants and zombie variations
- A PvZ Reviewer's Guide was created to help the press understand the depth of the game. The Reviewer's Guide detailed the additional levels of game complexity to counteract reviews that might conclude the game lacks depth. To augment the PvZ Reviewer's Guide, PR promoted the depth of the game by motivating press people to get to some particular achievement level. If they sent in a screenshot they got a particular PvZ patch.

PR event in the box: tchotchkes to convince the press that PvZ is a serious game and that PopCap is completely behind the game:
- Seed packets, energy drinks, magic bean plants, aprons, etc.
- Provide tools for evangelists to help them promote the game
- Avatar creator to create Zombie Facebook profile pictures
- Wallpapers/Buddy icons/Ringtones.

The Results

Success on PopCap.com and Steam led to a splashier launch into retail and additional game portals. The PvZ phenomenon was well under way within six months of launch and paved the way for retail distribution which started in August of 2008 and culminated with Halloween displays and third-party co-marketing and co-merchandizing.

To date, PvZ has received a positive reception from critics, garnering an aggregate score of 88/100 from Metacritic and an 89.5 percent from GameRankings. The review from Edge gave credit to PopCap for giving even the smallest details an extra touch of imagination, and for really putting their mark on the tower defense genre. Andy J Kolozsy (IGN) noted how much more content the game had compared with others in its genre and was impressed by how addictive it was, whilst Alice Liang (1UP) also found the game enjoyable and remarked how the lawnmowers guarding the screen's left edge added an extra dimension while still being easy to employ.

Chris Watters (GameSpot) focused on the design of the plants and zombies, in the course of praising the visuals generally and the game's overall value. The music video by Laura Shigihara received its own praise, Daemon Hatfield going as far as to say the video decided his interest in the game, while Alice Liang, rating the song, asked how on seeing the video anyone could not want PvZ?

1 — iPhone app developer

150 — Million copies sold

300,000 — Copies sold in first nine days

1 — Million USD gross sales in first nine days

#1 — In 20 top grossing country App Store charts

Canabalt

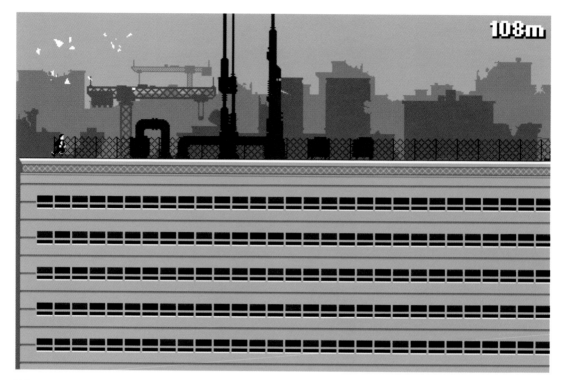

"This 'daring escape' platformer is simply
beautiful and beautifully simple."
Michael McWhertor, Kotaku

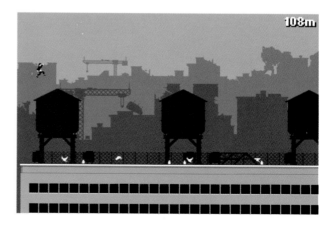

The Brief

The Experimental Gameplay Project was started by some now-famous game design students at Carnegie Mellon in the spring of 2005. The EGP is a game design activity where you make a whole game in just one week based around some kind of shared, central theme; sort of an exercise in creating with constraints. For August 2009, the EGP's theme was "minimalism". So Adam Saltsman ran with it.

The Challenge

The first challenge was building an interesting game under a very tight time constraint, an interesting theme constraint, and with a budget of zero dollars. Adam looked at old favorites for inspiration first – primarily the classic Super Mario Bros., but also lesser-known games like Flashback and Another World, which used novel and fluid game animation techniques to create believable heroes.

Instead of creating each level by hand, Adam made some basic-level building blocks, and wrote an algorithm to automatically arrange them in interesting ways. He also deliberately selected a visual style that was evocative and interesting, but simple and efficient to produce. Daniel Baranowsky composed the memorable music in just four hours, and for free. The result was one of the most popular Flash games of 2009.

Client
Semi Secret Software

Credits
Adam Saltsman, Eric Johnson,
Daniel Baranowsky

Awards
Best App Ever Award (2009 Best Productivity Killer); Mochi Award (Best Flash Game Port); Apple's Best of 2009 iPhone Games; Fantastic Arcade selection; Webby Nomination

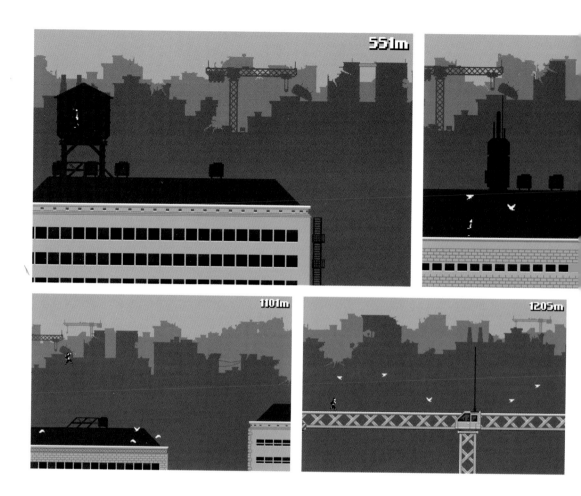

The Solution

Around this same time, Twitter was starting to get pretty big. Adam had tried doing traditional high-score systems in Flash games before, but they were too easy for abusive players to hack. They also take a bit of time to actually implement, and we were on a very tight schedule for this project. So, on a whim, we dropped in the ability to post players' high scores to Twitter, along with a brief message about how they died in the game, and a link back to the game. Canabalt was one of the first games on the web to do this, and it was huge.

A year earlier, in autumn 2008, we had a breakout iPhone hit with Eric Johnson's word puzzle game Wurdle. We'd been struggling for months on our sophomore title, an original arcade racing game, when Adam created Canabalt. After the first half million visitors hit the website during the first week, it became clear that we had something special on our hands. Eric dropped everything and ported Canabalt to iPhone in ten days, and we submitted it to the App Store.

The Results

Between the Flash game's popularity, our novel use of Twitter, and our existing iPhone customer base, we managed a very successful launch on the App Store without a publisher or backing from Apple. We were selling more than 2,000 copies a day right away, and we sold over a quarter of a million units on the App Store as of this submission. More than a year later, Canabalt is still helping to keep the lights on at Semi Secret Software.

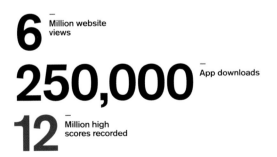

6 Million website views

250,000 App downloads

12 Million high scores recorded

Galaxy on Fire 2

"This is the most polished and professional game I've played on the iPhone. Great visuals, good voice acting and tons of gameplay. This is a high quality 'Eve'-like space combat and trading game of the very highest quality. Buy it and then get out there and explore the universe."
BadmanMonkey, UK user

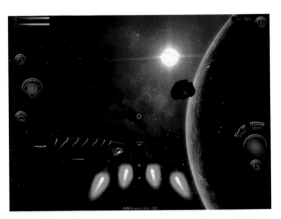

The Brief

We set out to make the deepest and most engaging space-trading sci-fi game that has ever been played on a smartphone. Bring new life to this classic genre and make it fit for the smartphone future. Produce a premium benchmark in the genre niche that serves as a major brand asset for the company. Deliver a game suitable for the broadest range of devices possible, performing at the respective maximum rate of the technical device characteristics. Position the title as the best game in its genre among users of all smartphone platforms. Market the title as highly lucrative content to the big players of the smartphone industry.

The Challenge

The project provided challenges in all of its aspects due to its sheer size. Creative design, sound design, graphic design – all of these required amounts of work never before put into the production of a smartphone game. Keeping the project alive while working on revenue makers at the same time was strenuous for the whole team, Fishlabs being an SME.

Keeping motivation high was a major challenge. After all, without passion and the commitment of each and everyone involved a game such as Galaxy on Fire 2 would not be what it turned out to be. Depth and size of the game presented us with ever newer issues in the year-long process.

Client
Fishlabs Entertainment GmbH

Credits
Fishlabs Entertainment GmbH
www.fishlabs.net
Music Composition
Periscope Studio
www.periscopestudio.de
Voice Acting
T-Recs Studios
www.trecsstudios.com
English Localisation
Pure Square Go
www.puresquaregogames.com

Awards
Slide to Play Must Have Award;
TouchGen Editor's Choice;
Pocket Gamer Silver Award

2.6 — Million game sessions in six weeks

3.9 — Million items equipped to ships

2.3 — Million missions assigned

50 — Million Kills

The Solution

We hired the right people to begin with. The team consisted of able and highly motivated people who know their craft. The challenge to create a truly epic game for the iPhone and other smartphones was presented to gaming aficionados who shared the same dream. We communicated our vision and kept it updated with project reality. Giving team members both responsibility for their field of work and the freedom to bring in their own ideas created ownership throughout the whole company. We had intense feedback sessions on all aspects affecting gameplay over and over again. Getting everyone on board in the striving to create a gaming experience unprecedented on a smartphone solved every one of our problems.

The Results

It is fair to say we have created the new benchmark for the sci-fi genre on the App Store with thousands of glowing fans worldwide that have been waiting for years for someone to bring the good old days of Elite, Wing Commander and Freelancer back to life on iPhone and iPad. Also the press shares the same kind of passion for Galaxy on Fire 2 underlined by 13 glowing reviews with a solid 8.8 average rating on the iPhone Quality Index Rating. Beyond that, some true fanatics have been playing Galaxy on Fire 2 for almost 100 hours which is even more amazing considering this is all on a mobile device.

The Pirata Boat Race

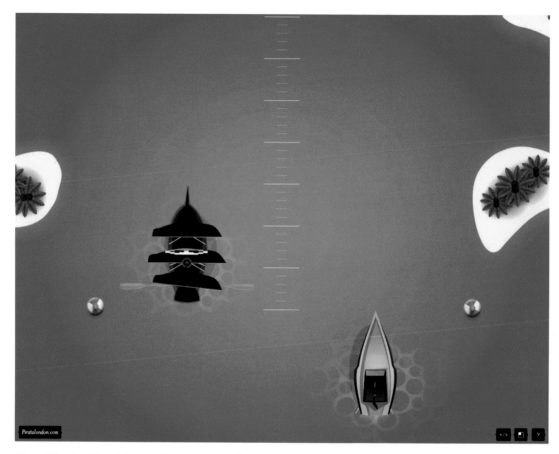

"The Pirata Boat Race is a computer game that uses your iPhone as a Wiimote. As you can see, it's also a sure way to look like a moron, but that's only half the fun of it."
Gizmodo.com

Client
Pirata London

Credits
Pirata London
http://piratalondon.com

Awards
FWA Mobile

The Brief

With increased Internet connection speeds and the now vast array of Internet devices becoming part of our everyday life, there is a whole new potential for creating more exciting and engaging user experiences. The Boat Race started out as an internal R&D project at Pirata. The brief was to explore the possibilities of cross-platform interaction, integrating the mobile and the desktop web environments via server-side technology. The aim was to come up with a fun game that did all of the above, ideally with a pirate theme.

The Challenge

Only a few developers are fully taking advantage of the iPhone's features, and only a small number of applications use it to integrate the device into a bigger experience beyond the handset's display, choosing instead to limit it to the confines of its case.

The challenge was to design a simple game that uses the device as a controller, and to figure out how to enable communication between the controller and game engine.

The obvious choice for this would have been Bluetooth for short-range communication between various devices. However, this presented a barrier to wider participation – as it would have required a bespoke set-up for each installation.

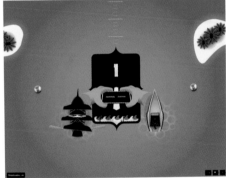

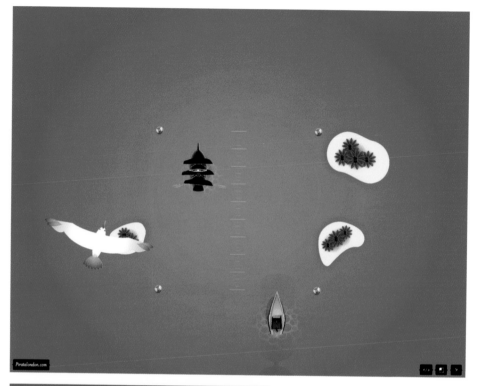

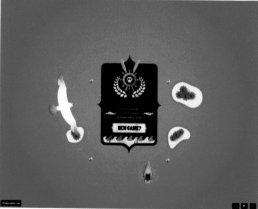

96,511 Games played
23,699 Downloads
47,100 Website visits in one month

The Solution

The solution, and the connecting factor, was to use the Internet. We designed and developed a desktop browser-based game in Flash, and simultaneously developed an accompanying iPhone controller application. The game itself acts as a hub, allowing multiple people to huddle around the screen, snap the QR code on their iPhones and propel their boat via the built-in accelerometer; a unique combination to provide a new and exciting gaming experience.

The hidden magic behind it all is the custom game server developed by us in PHP. Each time the game is run, the server issues a unique ID for each team, expressed as a QR code that the iPhone app decodes so that it knows which race session to join. Up to five players can join on each team simultaneously.

The Results

As development got closer to completion, and we realised how much fun it was to play, we decided to share it on the web. It quickly became very popular among the digital design community as well as much further afield. We received lots of player feedback, and even discovered a number of hilarious videos of members of the public competing in groups, including a bunch of guys at the Yahoo! Meme office.

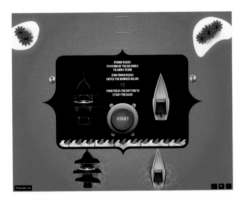

The Great
Niggle Nobble

"A simple fun packed idea, that proves
intelligent game play really helps a brand
engage over and above normal expectations.
Integrating the mobile app with the physical
world adds to its addictive nature."
Simon Gill, LBi

Client
O2

Credits
Agency Republic
www.agencyrepublic.com

Awards
FWA Mobile;
IAB Creative Showcase;
Campaign BIG Awards 10
– Finalist

The Brief

O2 set out to recognise the frustrations of thousands of home broadband users in their new animated campaign. They listened to consumers' niggles and wanted to show that they are determined to champion these issues on behalf of customers and deliver an end-to-end service that addressed their concerns. O2 characterised these broadband niggles in a fun and engaging multi-channel campaign, highlighting the lengths they were going to to nobble them. Our main brief was to create a fun, social and immersive digital campaign that continued to engage the audience with the Niggles and Narks characters beyond the TV and print.

The Challenge

We wanted to give our target market (mainly 25 to 35 year-old tech-savvy males) an engaging and social brand experience that was an escape from their day-to-day routine. With online gaming and Facebook being the core of their downtime we decided to create an addictive gaming mechanic that would encourage them to return to play again and again. With that in mind we designed the campaign messaging to appear around the game, not be the game.

The Solution

The Great Niggle Nobble was an innovative, fun and addictive Facebook and iPhone game that challenged the UK to rid the nation of broadband niggles.

The Facebook Flash game had players searching Google Maps and nobbling niggles into O2 stores with the aim of clearing the nation. The iPhone game brought a completely new way of gaming to the mobile device. Using the phone's compass and GPS we could show niggles hiding in the real world and reveal them in augmented reality.

Moving your iPhone around showed niggles lurking on street corners, running riot in parks or jumping on your laptop: you could even walk up to them to get a closer look. The iPhone game allowed you to nobble niggles on the go, boosting your Facebook score any time, anywhere.

The Results

In a six-week period over 20,000 players joined The Great Niggle Nobble on Facebook with a further 22,000 people nobbling niggles with the iPhone game.

Apple iTunes App Store: the game featured in Top 100 Free Apps (all categories) and Top 40 Games Apps. All in all well over 1,500,000 niggles were nobbled.

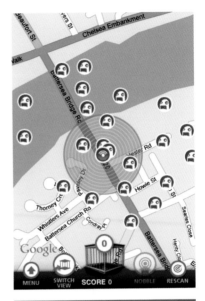

1.5
Million niggles nobbled

20,000
Players on Facebook

22,000
Players on iPhone

Mega Jump

"Mega Jump is the best casual
endless game in the App Store
that should be on every iDevice."
App Advice

Client
Get Set Games

Credits
Get Set Games
www.getsetgames.com

Awards
FWA Mobile

The Brief

Get Set Games started making games in Toronto, Canada in July of 2009. By spring of 2010 our indie studio had released two titles for the iPhone and iPad which, while being critically well received, did not produce the revenue we were hoping for. The first two games were innovative concepts somewhat outside of the well-known game genres and so found themselves in a small niche corner of the iOS App market. Both games had taken several months to produce because of the technical challenges we had to overcome. We wanted to try a new approach to both game design and production – a design that would appeal to a large proportion of iOS gamers and a production schedule that could be completed in under a month.

The Challenge

We like games that make us smile, we love games that make us laugh. The team at Get Set had enjoyed success working on smaller non-retail game projects, creating designs with lovable characters getting up to silly stuff.

We could tell we were on to a good thing if we broke out laughing while creating the game and we wanted to bring that feeling into our next game. We challenged ourselves to create a game concept and a set of characters that could make us laugh as we played, that was simple enough to be picked up and played by almost anyone, and that we could complete on a four-week schedule.

The Solution

The most accessible games are simple to play and have a control scheme that seems invisible to the user. Our game wouldn't need instructions to play it, and gamers would be able to identify with the main character as soon as they began. Mega Jump was created to build on these concepts and expand on some of the popular new game mechanics that had evolved in the iOS App market. The game would be played by simply tilting the iOS device from side to side – no buttons or controllers. We chose Redford, an adorable little red monster with huge eyes and a toothy grin, to be the star of the show. He was originally designed and created for a two-day gaming event but soon became the mascot of our company. Redford would be put front and centre of the game and marketing materials because just looking at the little guy made us chuckle and we hoped he would do the same for other gamers. Mega Jump would be vibrant and fast-paced, bursting with humorous sound effects, full of bright shiny stuff for the player to collect on their journey, and stacked with exciting and funny surprises to keep the player engaged.

The Results

Mega Jump has been an incredible success for Get Set Games. It was released at the beginning of May 2010 and by the end of the month it had been downloaded by over a million people. We now have over 7.5 million players around the world and that number continues to grow at an amazing pace. Around 400,000 people play the game every day for an average of 40 minutes each. A million games of Mega Jump are played globally each day. Mega Jump became the number one ranked App in Apple's iTunes Store in 38 countries and the number one Game in 50 countries.

Mega Jump has been updated 11 times in the seven months since release with major new content, characters, and gameplay advances to keep players coming back for more. We now enjoy an active community of gamers who talk about their experiences with Mega Jump and its characters through our website, Facebook, Twitter, and YouTube channels. Mega Jump has put Get Set Games on the map and created a bright future for our company.

7.5 Million downloads

122 Million games played

1 Million games played per day

400,000 Unique players per day

Trainyard

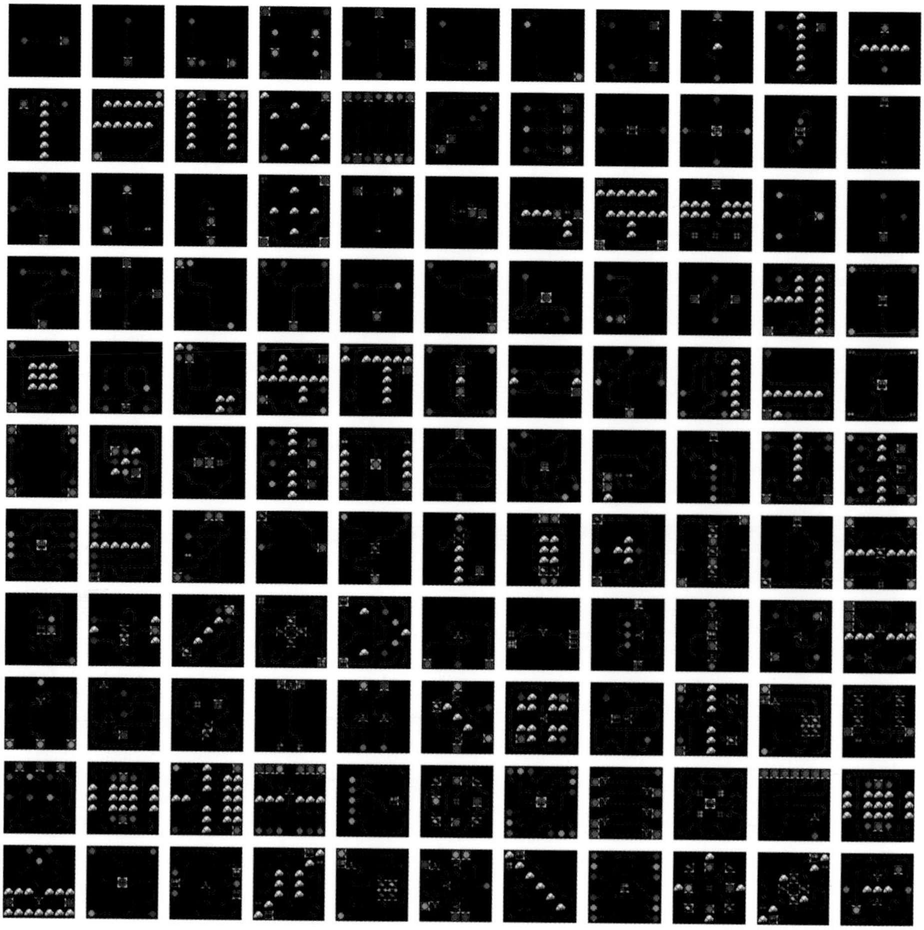

"This is one of the must-have actual 'puzzle' games on the iPhone. None of that block-dropping, gem-matching, ball-shooting or word-making that clogs up the App Store. Just thinking power versus an honest-to-goodness brainteaser. I love it."
Skamando, App Store user review

Client
Matt Rix

Credits
Matt Rix
Blog: http://struct.ca
Twitter: http://twitter.co./MattRix

Awards
FWA Mobile
Chomp's #1

The Brief

I had a day job as a Flash developer for a digital advertising agency, but I wanted to do something in my spare time to work towards my goal of being a full-time game developer. Throughout my life I've enjoyed working on small games, but I hadn't actually released a game for a number of years. My hope was that I could create a game that would earn enough money to allow me to work on games from that point onward. I knew that my experience as a Flash developer had provided me with the right technical skills and usability knowledge to create a great game.

The Challenge

I decided to create a game for Apple's App Store. The App Store is an incredibly oversaturated market, yet I felt it was still possible to succeed if I created the right game.

It would need to appeal to a wide audience, but still have depth and unique gameplay in order to stand out from the crowd and get noticed. I had a budget of around $1,000 CAD, so it would also need to have a strong social component to help the game grow virally without any money spent on advertising. The game would also have to be simple enough that I could build it in the small chunks of time I had available.

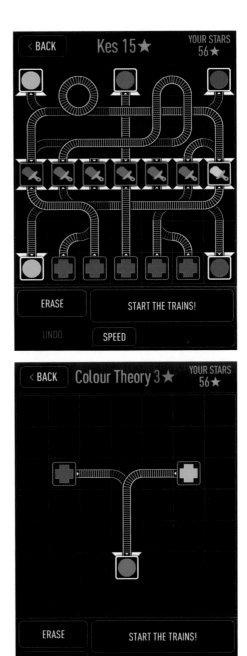

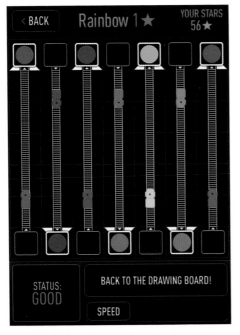

3.6 — Million express players

1.1 — Million solutions shared

92 — GB Solution images stored

15,000 — User reviews

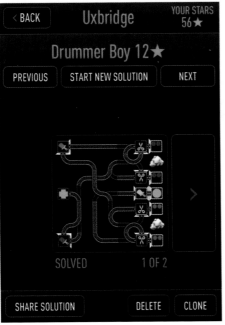

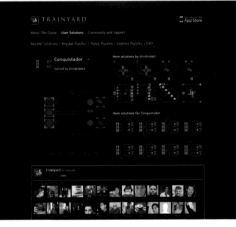

The Solution

I came up with the concept for Trainyard, a puzzle-solving game, and brought it to life over the course of a year. The core "track drawing" gameplay mechanic was unique, yet accessible, and the game had a vast amount of depth at the later stages. I taught myself how to program for the iPhone with Objective-C, and also did all of the design, sounds, and back-end programming by myself. On the social side of things, I gave users the ability to share their puzzle solutions online at http://trainyard.ca/solutions. To help market the game, I created a trailer and shared it with numerous iPhone app review sites.

The Results

The game's first four months on the App Store were spent in relative obscurity until I created a free "lite" version, which caused the game to be featured by Apple. It shot up the charts and reached as high as number two in the App Store's paid apps chart, even ahead of App Store heavyweight Angry Birds. The free version was downloaded over 3.6 million times, and the paid version sold hundreds of thousands of units.

The social sharing system was also a success. In total, over 1.1 million puzzle solutions were shared on http://trainyard.ca. The game brought in enough money to allow me to start my own game company and finally pursue my dream of creating games full-time.

Barclaycard
Rollercoaster Extreme

"Ok ok this is an awesome application now that they do the ghost racing I can barely take my eyes off this game."
<u>Michael De Miranda</u>, App Store user review

The Brief

Barclaycard had a strong heritage among older consumers, but amongst a younger audience, relationships with credit card providers tend to be distant and rational to the point where cards are simply viewed as a functional commodity.

In the '00s, brands like Egg and Smile successfully positioned themselves against well-known brands like Barclaycard. Egg and Smile's branding disrupted category norms and for a while successfully established them as the most innovative and relevant brands in this category. Our challenge was to improve engagement with Barclaycard and boost its image among affluent young audiences. With this in mind, we set out to radically improve its credibility and rebuild its status as a truly innovative brand.

The Challenge

Barclaycard's existing advertising content (Waterslide and Rollercoaster TVCs) and sponsorship properties (such as the Mercury Music Prize and Wireless Festival) gave us strong starting points from which to deliver engaging digital media. But in order to be truly credible, and force re-evaluation amongst target audiences, we needed to come up with a digital entertainment idea that would dramatically redefine what a financial services brand could be. So we decided to build a market-leading gaming brand, creating a franchise of iPhone games based on Barclaycard's advertising activity.

Client
Barclaycard

Credits
Dare
www.daredigital.com

Awards
Top 10 Apps 2010
(Campaign Annual 2010)

The Solution

Throughout the activity, our approach was to produce games that didn't seem to be afterthoughts, but which were genuine, high-quality entertainment in their own right. We wanted to produce iPhone games which could stand alongside the best games already available for the platform. Both games take the "glide through life" concept and visual themes of BBH's Waterslide and Rollercoaster TVC advertising and put our players right in the heart of the action they'd seen on TV. Waterslide Extreme employs the iPhone's accelerometer to create an intuitive interface through which players can guide themselves along a fantastic aquatic joyride.

Like all great sequels, Rollercoaster Extreme expands on its predecessor, creating a more involved gaming experience to encourage longer and repeated play. Players ride the rollercoaster through four detailed New York environments, racing the clock and avoiding obstacles to rack up points and trophies. Updates of the game expanded on this further by letting players race each other – and allowing iPad owners to download one of the first dedicated branded games for the device (both August 2010).

Both games were seeded to the gaming community prior to release using demo videos. While Waterslide was a story of innovation in itself, being one of the first branded games for the iPhone, Rollercoaster was launched into a much more established and competitive brand and gaming environment. In common with major gaming franchises, we decided to create a buzz around the launch and encourage ongoing play and advocacy.

The Results

Rollercoaster Extreme has been downloaded ten million times to date, at a faster rate than its predecessor, despite a more competitive store environment.

Within 28 days, Rollercoaster Extreme had been mentioned in 9,400 blogs. 900,000 players have visited the Barclaycard.com site from Rollercoaster Extreme.

Among Players of Rollercoaster Extreme in Barclaycard target audiences (which comprise 70 percent of iPhone ownership according to TGI), Dynamic Logic recorded increases in Awareness and recall metrics (Message Association, Online Ad Awareness, Brand Favourability, Brand Love), specific service related messages ("Is innovative", "Is on my side", "Is changing the way people pay for things") and Purchase Intent.

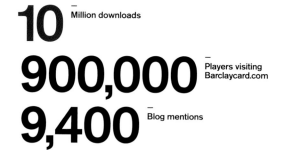

10 Million downloads

900,000 Players visiting Barclaycard.com

9,400 Blog mentions

The Cube

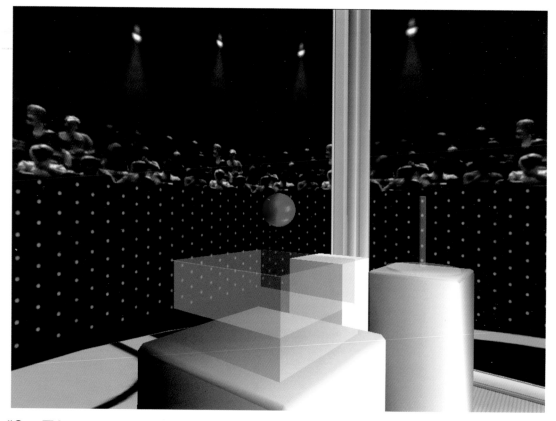

"Our TV team worked closely with our developers to create a unique TV show app that looks great but also plays extremely well as a game, avoiding the common trap of creating a great-looking game which is no fun to play or vice versa."
Adam Adler, The Cube format creator

Client
All3Media International Ltd

Credits
Swipe Entertainment Ltd
www.swipeentertainment.com

Awards
UK App Store: Number 1 Grossing App; Number 1 iPad App; Number 3 App Overall; Number 1 Arcade App; Number 1 Family App

The Brief

Recreate The Cube experience for iPhone/
iPad and allow players to take on The Cube
challenges in a realistic recreation of the
TV set and show experience. The game may
seem simple, but under the studio lights,
with a hushed audience and a substantial
prize sum at stake, an apparently easy task
becomes a dramatic test of nerve. Design
challenges to push users to the limit of their
skill. Recreate the tension of the show and
increase the difficulty and tension as the
prizes increases. Create a game that offers
premium values, extended gameplay
potential and variety.

The Challenge

We wanted to keep everything as close
to the original TV show as possible whilst
also showcasing all of the unique abilities
of the iPhone and iPad. This meant that each
game required a completely different control
treatment. In addition, the TV show uses
music scored for each specific attempt
and contestant. Mirroring this within the
constraints of older iPhone hardware was
also something of a challenge.

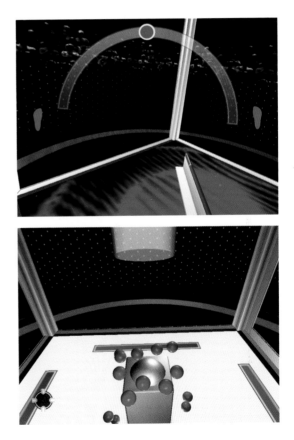

The Solution

Due to the nature of The Cube it was logical to recreate a true 3D representation of the set. The Cube in the game is an inch-perfect scale model of The Cube in the studio. We then employed realistic movement and physics in the game to portray a feeling of actually being inside it.

We carefully translated 15 of the real gameplay challenges from the show to challenge the user. Many with totally different interaction and gameplay mechanics.

In order to completely immerse the player we used real assets from the actual television show to ensure that the sound and visual experience was completely authentic.

The Results

The game reached the number one position in the UK iTunes App Store (Top Grossing Apps) and consistently appeared in the top five over the following six to eight weeks. It was played over 2.5 million times in the first eight weeks alone and has had over 300,000 profile registrations so far. There are over 22,000 lines of game logic code and the lead programmer consumed getting on for 400 mugs of coffee! We also managed to break a server on launch night.

2.5
Million games
played (in 8 weeks)

300,000
Profiles registered

22,000
Lines of game
logic code

384.5
Mugs of coffee
consumer by lead
programmer

1
Server broken on
launch night

Flick Football

"This game is very very addictive!
First it starts with hours, then
days, then to weeks! Heeeeelp!!
I need to sleep! I need to get up!
I need to woooork!!"
App Store user review

Client
Neon Play

Credits
Neon Play
www.neonplay.com

Awards
App of the Day with Pocket-lint

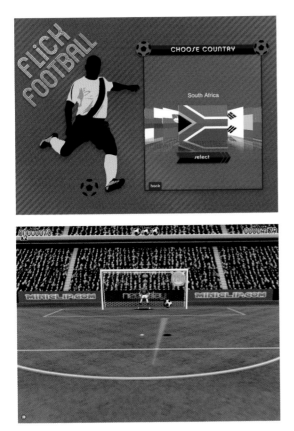

The Brief

Our aim was to create the most beautiful, playable, fun, simple but addictive casual football game on the iPhone. In the run-up to the World Cup, we identified that there didn't seem to be any really playable and fun free-kick football games on the App Store. So we set out to create the best game we could, but we only had six weeks as we were a new mobile games studio and this was our first app. The World Cup was approaching, we didn't know how long Apple would take to approve it and we would be going into uncharted territories with every single element of the game build. This was to be a steep learning curve…

The Challenge

Our real challenge was how to actually make an app. The idea clearly came first, so we researched to check there was nothing similar on the App Store, which there wasn't. Then we wireframed (vaguely) the flow and structure of the app (which changed a lot as we went along). Next we had to design a standout game that would resonate with the audience. So we needed a 3D stadium, a wall of players, a diving goalkeeper and realistic ball movement.

Also, we needed viral mechanics to make the game competitive and addictive. The biggest challenge though was how do you make a football game stand out on the App Store, when there are already 250,000 apps out there? A partnership with Miniclip.com helped get the word out there.

The Solution

We had to put our money where our mouth was and make the most addictive casual football game on the App Store (we hoped!). The look and feel was so important and after lots of research and thinking, we took the 1970 Mexico World Cup logo as our starting point and the design evolved from there. We commissioned a 3D artist to create a Mexico-style football stadium and to create a wall of animated defenders and a diving goalkeeper.

The developer set about creating a flick mechanic that was both logical, fun and easy to control – testing this relentlessly was imperative and we eventually completed what we felt was a really compelling, fun and addictive game. Plus we needed game mechanics that kept people playing with global and Facebook leaderboards. We learned a lot…

The Results

We had a free and paid version of the game and together they have had over one million downloads since June 2010. Not surprisingly, there are more free than paid downloads, but the upsell from free to paid was good.

The game has remained in the Top 40 UK games for over six months and has had success globally but with the UK being 50 percent of downloads. Apple really supported the game and it featured in the Soccer sections on iTunes/App Store during and after the World Cup. It's a four-star rated game on the App Store and its success seems to be long-lasting.

1 Million plus
downloads

556,180 UK users

5.6 Million game
sessions

163.5 Million balls kicked

2.7 Million practice
sessions

Monster Dash

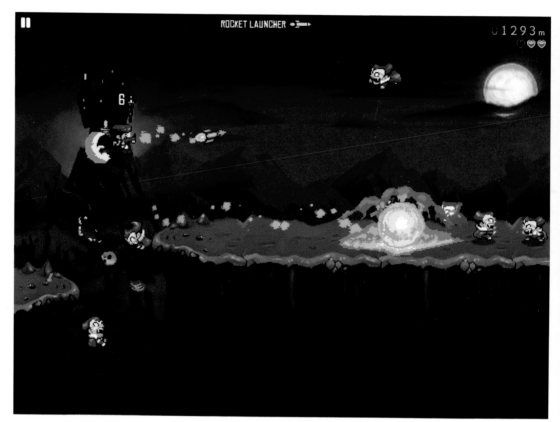

"It combines gorgeous art with perfect controls and just enough innovation in the 'wallrunning' iPhone genre to earn a place at the top of your game list."
Mike Schramm, TUAW

The Brief

While Fruit Ninja was busy being a runaway success on the App Store, there was the hefty task of following it up with a new game. Our plans? Improve upon greatness.

Fruit Ninja was an extremely original and unique title when it was released, but we didn't necessarily need to keep matching and exceeding every characteristic with every new game. We can pick and choose our battles; pick and choose where to excel and where to strike the right tone with a new game. Most of the most successful games today are strongly rooted in a conventional genre, and the iPhone features a strong suite of quite unique gameplay conventions. We took on the endless runner.

The Challenge

In general, running games are one-button affairs that involve jumping over obstacles at increasing speed. We could create a game like that in a day, but to really make a mark in the format we needed some new tricks.

We knew pretty much right away that we wanted to add a second button for combat. It was a natural progression of the genre, and even though we were essentially doubling the input requirements for the player, it was a necessary step to take it to the next level.

We were also tossing about names, and although it only featured zombies at first, to maximise options for updates and unique settings, we went for a general monster theme – and are very happy with the decision! Monster Dash was born.

Client
Halfbrick

Credits
Halfbrick
www.halfbrick.com

Awards
Pocket Gamer

400,000
Copies sold

62
Awards to unlock

1
Million plus
kilometres run

5
Costumes hired to
create trailer

The Solution

We needed a great character to add some personality to the action, and found one right under our nose. Barry Steakfries was the hero of our earlier game Age of Zombies, and we discovered that his witty remarks were a perfect fit for the light-hearted action!

To add even more depth, we created multiple weapons to keep the action fresh, and unique levels which Barry transports to at regular intervals. With this tech in place, it was easy to update with new levels, new weapons and new items, something which many other running games neglect.

At a certain point we took a broad look at the game and knew that it was ready. We reached our innovation benchmark and delivered a simple, polished title which was easy to play and immediately recognisable to iPhone users.

The Results

By expanding and improving upon the marketing techniques used with Fruit Ninja, in August 2010 Monster Dash enjoyed one of the most successful iPhone releases of all time. In just one day it shot to the top five charts all around the world – and actually occupied a space right next to Fruit Ninja in many cases, when Fruit Ninja was still a major success!

However, Monster Dash didn't have the extreme level of addictiveness and viral stickiness that Fruit Ninja did, and after a couple of weeks sales were waning. As Halfbrick became known as a top-tier developer around the world, and as Fruit Ninja continued to sell massive numbers, Monster Dash established a snug spot and became a great, profitable addition to our portfolio.

Hungry Shark

"Hungry Shark is a cute, fun, simple game
that's pretty hard to put down. It has a
just-one-more-time appeal that I would
liken to Doodle Jump; it seems that surely
the next game will win you a crown spot
on the global leaderboards."
Touch Arcade

Client
Future Games of London

Credits
Future Games of London
www.futuregamesoflondon.com

Awards
#1 in Japan App Store

The Brief
Hungry Shark was the debut title from Future Games of London. As such, we were setting out our stall as to what we thought a great mobile game should be like: a simple idea, flawlessly executed, then well supported with updates taking into account user feedback. We had six months' worth of cash. In that time we'd have to develop the game, publish it and earn back our development budget or there'd be no future games from us.

The Challenge
One challenge we faced was to complete the game before our startup capital ran out! The team had all worked together before at Shadow Light Games and the concept was pretty well understood. So we were confident we could make the product. We'd taken the game idea to several publishers in the past who declined to fund the project, now was our chance to prove them wrong. Having never published a game on our own before, this was going to be interesting. We had zero marketing budget, so everything relied on good user feedback and getting the word out through websites and most importantly – convincing Apple that they should promote the product for us. A huge risk. But if self-publishing was going to work, it had to be done.

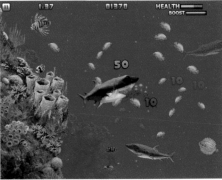

The Solution
Everything relied on the game being awesome. With no licence to rely on and no marketing budget, this game was going to sink or swim on merit only. We looked at every popular game on iPhone for clues as to what people liked about these games. We identified that the name and icon were crucial – these determined if a customer would look any futher. Furthermore, the game had to look great and be rewarding to play from the beginning. Features and volume of content were pretty much irrelevant. Core gameplay done well was what mattered. We play-tested the game to death, continually making tweaks to improve the experience. We avoided the temptation of 3D graphics and technical wizardry in order to concentrate on the gameplay, which we believed was the key to hitting the mass market. We honed and polished the product until we could delay no more.

The Results
Hungry Shark sold well beyond our expectations. Initially we failed to get visibility in the US, the largest single market, but spurred on by strong sales in all other major territories we invested time in producing part two, and then in a crucial and brave move we released part one for free. We were convinced that if only US iPhone owners knew about the product, they'd buy it in droves. The results were spectacular. Practically overnight we went from selling 1,000 units a day to a peak of 15,000. As luck would have it, just as we released the free version, it was Shark Week in the US, an annual Discovery Channel event. As one commentator remarked "an inexplicable shark fad is sweeping the nation". Thank you Discovery Channel!

650,000
Paid downloads

7
Million free downloads

1
Million YouTube views

48
Countries voted free version #1 in App Store

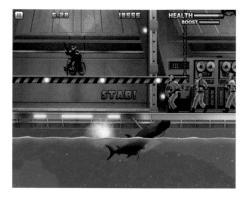

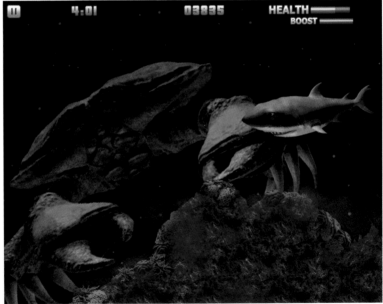

Antrim Escape 1&2

"Antrim Escape is very, very easily my favourite game on the Store. When I wasn't playing it, I was thinking about it! I'm proud to say I finished it without a guide and while it was very hard, the reward of knowing I'd used my mind was so worthwhile that I'd recommend everybody give it a go!"
Chris Catterall, user

Client
Game Hive
www.gamehivegames.com

Awards
FWA Mobile; featured by Apple
in "New and Noteworthy" and
"What's Hot" category

The Brief

Antrim Escape is a room escape game. The objective of the game is to find, combine and use designated items to solve puzzles and eventually find the exit of the room. In order to succeed in the game, you have to comprehend the functionality of the iPhone (multi-touch, accelerometer… etc.). With its realistic graphics, enticing storyline, intellect puzzles and intuitive controls, it is a brand-new experience compared to your previously played room escape games! The game will take you days or even weeks to solve the puzzles.

The story is based on a haunted garden in Northern Ireland!

The Challenge

At the time of developing Antrim Escape, App Store had already been crowded by all sorts of popular games produced by the big companies. As indie developers, our biggest challenge is to find out the kind of game that has not been done well enough on iOS and make it a whole lot better. We know that we will never make a better racing game than Need For Speed, but there are definitely some niche games where we can make impacts. The iPhone has a unique multi-touch screen and accelerometer that we can take advantage of, but also lots of drawbacks such as small screen size, limited RAM and unexpected call interruptions. How do we find the games that would appeal to the mass iPhone markets?

1.4
Million downloads,
Antrim Escape 1

20,000
Downloads of
Antrim Escape 2
in first 2 weeks

#58
Paid overall apps
in US

#30
Paid overall apps
in UK

4.5
Stars average rating
in 300 reviews
worldwide

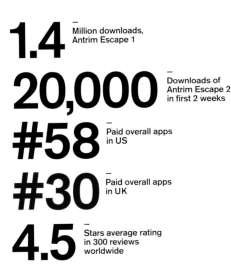

The Solution
We started with brainstorming a list of games that we love, and evaluated the competitor landscape of each game. Unlike many other game types, room escape games are represented poorly in a couple of titles. We felt that the success of room escape games on other platforms was the evidence provided by a possible large customer base, so we decided to make a room escape puzzle game designed for iOS. We also wanted to open people's eyes by using many unique iPhone features that have not been used on PCs or other mobile consoles. We further optimized the game for iOS such as zoom-in, graphics memory management, and auto-save to address iPhone-specific problems.

The Results
The result was a room escape game that was not like any of the existing ones. It has iPhone specific puzzles, a full story line, realistic graphics, four available languages and is easy to play on a small screen. The reviews were wildly positive and regarded it as the most original and brilliant room escape game on the iPhone.

In the successor, Antrim Escape 2, we introduced new innovations such as 3D puzzles and team-tactic puzzles to enhance the overall gameplay even further. The Antrim Escape has become a franchise that is well known in the iOS world as the premium puzzle game.

M-Commerce

Introduction by
Ian Wharton, Zolmo

02

M-Commerce M-Commerce M-Commerce M-Commerce

Here's something I have heard a lot over the past few years: "Mobile commerce is still in its infancy." The reality is m-commerce has been around for some time, since 1997 to be exact. A more fitting description of the industry would be a largely misunderstood but ambitious teenager that will spend the next five years trying to define itself by way of experimentation.

"'Mobile commerce is still in its infancy.' The reality is m-commerce has been around for some time, since 1997 to be exact."

There is no doubt m-commerce (or whatever alias it will ultimately attribute to itself) is one of the fastest-growing global industries. Consider briefly the potential: mobile phones are more prevalent than credit cards, most of us are never more than a metre away from them, even at night, and the uptake of smartphones in 2010 looks like a hockey stick. There is no shortage of technology and there's certainly no shortage of content to deliver great commercial experiences. So why do I write this introduction so tentatively? It all comes down to a question of perceived value.

Understanding digital value

We have spent the better part of two decades educating ourselves in the commercial mechanics of the Internet. Purchase physical products online and they show up at your door three days later. Simple. However, it is the mobile industry that has stuck its head above the parapet to let consumers explore something revolutionary – the value of digital products.

The App phenomenon underlines the commercial prospects of mobile devices and things look more than promising. Downloads via Apple were set to pass the ten billion mark in early 2011 with over $1 billion already paid out to developers. It's a level playing field and anything is possible, but unfortunately there is one thing holding developers and publishers back. Consumers currently have a skewed appreciation of the worth of digital goods and this is partly down to bad habits picked up from the web.

Throughout the digital age we have curiously been giving all our premium content away for free on the Internet, news being a prime example. The ability to monetise this content has only suddenly become viable through mobile applications due to the simplicity of the buying process. One-click purchasing through a single App Store account and the content is on a device that's with you 24/7. These are uncharted waters but thanks to pioneering news companies like The Guardian and Financial Times exploring new ways of delivering their content digitally, mobile users are growing accustomed to paying for it. Just how significant this is becomes apparent when you consider how we have come full-circle.

"The most successful commercial ventures on mobile platforms have not been throw-away products but considered businesses with long-term plans."

One could argue mobile commerce has influenced the nature of the web as we see more and more editorial content now sitting behind pay walls. It has taken the commercial fortitude of mobile to get the ball rolling, and users are taking the first steps to entertaining the idea of digital content being as valuable (if not more so) as its physical counterpart. Quite right, why should it cost less than the one on the news stand? Once consumers embrace the financial models that become sustainable for the developers – such as Freemium – things will get very interesting, very quickly.

Build businesses, not gimmicks
It's not all down to the consumer, the future of mobile commerce rests equally on how brands and developers approach apps. The most successful commercial ventures on mobile platforms I have seen have not been throw-away products but considered businesses with long-term plans. Brand-building, marketing, iterative development and intelligent fostering of exciting partnerships need to be just as important to a software developer as they are to Nike when they design a new sportswear range. Developers need to be crazy enough to think their product can change the world.

Money revolution

Apps aside, something you are likely to hear a lot in 2011 is Mobile Money. If you picked up this book in 2015 you may well have paid for it using your mobile phone. That's the kind of big-picture thinking that is driving mobile commerce forward. Never before has a platform been so social and so ubiquitous that it can champion this kind of revolution. The reality is perhaps closer than we would imagine and when this technology diffuses into the hands of millions it will have a profound impact on our lives.

"New financial models will always bring about confused and disgruntled users with one-star reviews but the fact remains, if you infuriate a handful of people you are probably doing something right."

Whether it be making payments at retailers, sending funds and apps between one another or receiving offers based on proximity to our favourite stores, the mobile industry is poised to fundamentally redefine the way we handle transactions. The challenge for everyone in the industry is to engage consumers with technology that is being redefined more frequently than anything that has ever preceded it.

Think big, make something inspirational

As you'll see from the case studies in the chapter, the most successful m-commerce projects are not necessarily the first to market – nor are they always apps. They are the projects that aim to define genres, flex the potential of mobile platforms by doing something ground-breaking and by offering the best experience. There's no room to be cautious in an industry with so much potential. Be cavalier. Dare to fail. Sure, new financial models will always bring about confused and disgruntled users with one-star reviews but the fact remains, if you infuriate a handful of people you are probably doing something right. And that's how teenagers should get their kicks.

Ian Wharton
Zolmo

Bio.
Ian Wharton
Zolmo

Ian Wharton is the Creative Director and Partner of award-winning software publishing company Zolmo. An Art Directors Club Young Guns 8 winner and Director of an acclaimed short animated film, he has since worked with some of the world's top brands and creative teams.

Previously an Art Director at leading visual effects company The Mill, he designed commercials for Audi, Nokia, EA and game trailers for Sony. At 23 he was appointed Associate Creative Director and led the company's global rebrand across London, New York and Los Angeles.

In 2009 London-based Zolmo partnered with Jamie Oliver to design and develop the top-grossing and top-rated app 20 Minute Meals. After just two years Zolmo is already considered a pioneer of mobile publishing and has recently been featured in the D&AD annual and honoured with the prestigious Apple Design Award.
–
www.zolmo.com
www.ianwharton.com
@ianwharton

"Developers need to be crazy enough to think their product can change the world."

Jamie's 20 Minute Meals

"The daddy of all food apps."
The Observer

The Brief

One of the world's best-loved celebrity chefs Jamie Oliver teamed up with software publishing company Zolmo with a goal to bring great cooking to people short on time. Jamie wanted to give millions of people the tools they need to help them create great-tasting, homemade meals in 20 minutes – less time than having a takeaway delivered. The iPhone was the perfect platform to deliver this message to people on the go.

The Challenge

To create a product that truly works for the medium and transform the iPhone into the perfect cooking companion. The app needed to have variety, be accessible and packaged into a beautiful interface like nothing people had ever seen before. From the project outset we were very mindful not to simply re-purpose the cookbook but to create an entirely new way of cooking. In doing so we hoped to tap into a new demographic of would-be cooks: those most attached to their phones but who had not yet been swayed by books or TV. Most importantly, we wanted to bring all the great content Jamie had put together into a product that was a joy to use.

Client
Jamie Oliver

Credits
Zolmo
www.zolmo.com
Jamie Oliver
www.jamieoliver.com

Awards
FWA Mobile; Apple Design Award; D&AD 2010 Annual; iTunes App of 2009

#1 — Top grossing app

#1 — Top-rated lifestyle app

The Solution
The app features several world-firsts. 250 exclusive recipes purpose-written in step-by-step form, each one accompanied by beautiful photography and handy voice prompts from Jamie. Over an hour's worth of video hints and tips from Jamie and a powerful search by ingredients or recipe. There is also an interactive shopping list which conveniently sorts by supermarket aisle for speedier selection in-store. If you are unsure what to cook, the app will suggest a recipe for you with a quick shake of the phone. Cooking has never been easier.

The Results
20 Minute Meals has been featured in Apple television adverts in both the UK and US. It has been app of the week and named iTunes app of 2009. 20 Minute Meals has also led the charts in top-grossing overall and top-rated in the Lifestyle category. One year on the app is still a favourite among a collection of cooking apps and has been honoured with the prestigious Apple Design Award.

The Elements:
A Visual Exploration

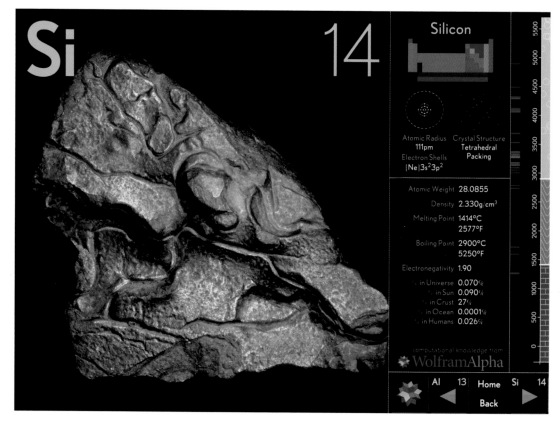

	Silicon
Atomic Radius	Crystal Structure
111pm	Tetrahedral
Electron Shells	Packing
$[Ne]3s^2 3p^2$	
Atomic Weight	28.0855
Density	$2.330 g/cm^3$
Melting Point	1414°C
	2577°F
Boiling Point	2900°C
	5250°F
Electronegativity	1.90
% in Universe	0.070%
% in Sun	0.090%
% in Crust	27%
% in Ocean	0.0001%
% in Humans	0.026%

computational knowledge from
WolframAlpha

Al 13   Home   Si 14
◄   Back   ►

"Best App of all Theodore Gray Wolfram Periodic Table. Everything is animated and gorgeous. Alone worth iPad."
Stephen Fry on Twitter 3/31/2010

Client
Touch Press

Publisher
Touch Press LLP

Credits
Writer/Creator
Theodore Gray
Photography
Theodore Gray and Nick Mann
Project Management
Max Whitby
Software Development
John Cromie

Awards
Chicago Book Club, Award of
Excellence and Best in Show;
Brit Insurance Designs of the Year
(nominee Best Interactive App);
Apple iTune App Hall of Fame;
Gizmodo's Essential iPad Apps;
The Telegraph: 5 of the Best Apple
iPad Apps; US News & World Report:
Top Ten Best iPad Apps for College
Students; Huffington Post: Eleven
Bestselling Book Apps for Adults and
Kids; Society of Publication
Designers: 10 Essential iPad Apps for
Publication Designers; App Storm:
30 Examples of Stunning
iPad App Interface Design

The Brief
When Steve Jobs announced the iPad in
January 2010, with availability in April, we
knew we had 60 days to take our website
(periodictable.com) and our existing book
(*The Elements* by Theodore Gray) and create
a piece of software that would do justice
to Apple's revolutionary tablet. We wanted to
design an electronic book that would realise
the true potential of the new interactive medium,
while fully respecting all the vital characteristics
that make printed books such an effective and
enduring form of communication. The app
needed to be highly responsive, to take
advantage of the special features of the
platform and above all to be beautiful.

The Challenge
We faced three main challenges in creating
The Elements: A Visual Exploration. The first
was to produce a design that really takes
advantage of the iPad's superb high-resolution
multi-touch screen. Secondly, we needed to
find ways to use the device's excellent graphics
sub-system to mask relatively long load-time for
data-heavy photographic rotations that are at
the core of the book. Thirdly, we needed a
highly-automated production system capable
of handling tens of thousands of source files.

Each one of the 500 objects in the book
was photographed in 360 or 720 positions
on a precision turntable. There was no way
that conversion to the iPad could be
accomplished manually.

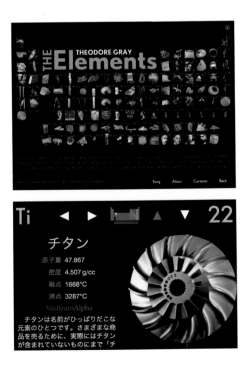

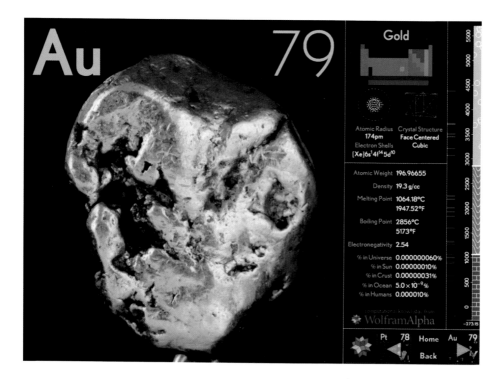

The Solution

We decided to follow the architecture of the printed books and give each element two pages. On the first page we included a beautiful animated rotation of one element sample along with facts and figures including a link to the Wolfram Alpha computable knowledge engine. This provides up-to-the-minute data including, for example, the current price of gold, platinum and silver. On the second page for each element are the rotations. At first sight they look just still photos in an illustrated book. But on the iPad you can reach out and spin each one.

It responds with silky smoothness, turning to show all sides. A quick flick and you discover the objects have "physics": they will continue to rotate for a period that depends on how hard you spun them. Loading in these data-heavy rotations takes an age in computing terms... up to a couple of seconds. To mask this delay we play a carefully configured animation in which the objects fall tumbling into place from the sides of the screen. The iPad can play these "spin-down" animations in its sleep (so to speak) while the hard work of loading in the rotations takes place in the background.

This design has the added advantage that the reader's attention is subtly alerted to the objects' interactive potential.

1

Million USD net revenue in first six months

150,000

Units sold in first nine months

60,000

Individual files required to compile executable

460,000

Google hits nine months after launch

The Results

The Elements: A Visual Exploration has been one of the most successful applications for the iPad. It has featured widely in Apple marketing including featuring for a while on the home page of apple.com and in many TV commercials around the world. It has been widely tweeted, reviewed and mentioned in blogs, attracting several thousand mentions in the press and regularly appearing in lists of the best apps for iPad. Inspired by this success we quickly translated the title into many languages including Japanese, French and German. Tom Lehrer's famous Element Song, which plays the first time the title is launched, was specially translated into Japanese, where it has acquired something of a cult status. The creators of The Elements: A Visual Exploration have gone on to establish Touch Press to develop and publish a wave of innovative new titles that apply the principle described here in other subject areas ranging from astronomy to poetry.

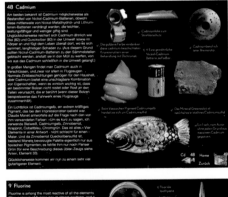

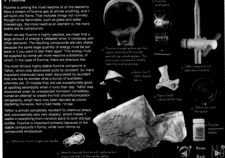

ASICS Australia Mobile Site

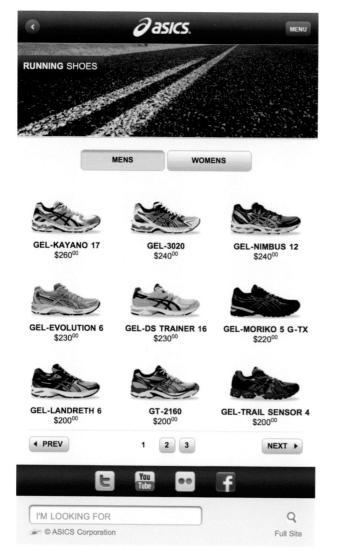

Client
ASICS Oceania

Credits
IdeaWorks Sydney (Y&R)
www.ideaworks.com.au

Awards
FWA Mobile; Kentico Site
of the Year

"The mobile site for us was a way to deliver content in a format that was dedicated to research on the move, or as we are finding more and more, research in-store. It enabled our customers to get the key information they needed, anywhere they were, including tools like our shoe finder, to help purchasing decisions at the POS."
Tara O'Malley, Online Specialist, ASICS Oceania

The Brief

In early 2010 IdeaWorks Sydney (Y&R) were tasked with developing a new, category-leading website for ASICS Australia, with part of that being a brief to create a mobile platform that would provide a rich product, research and shopping experience (supporting all smart phones) to complement the website's popular "send to phone" lead generation feature and provide an optimized experience for the increasing number of people using mobile devices on the move and in-store.

The Challenge

With the website delivering such a rich user experience, we had to find a way to extend that (and most of its tools) into a mobile interface that would provide a seamless journey across all our digital platforms.

The mobile site needed to deliver almost 90 percent of the full website's content in both vertical and horizontal formats, that was quick and easy to read, but could be expanded for users who wanted to explore product information in depth, ensuring the mobile site could be used to make purchasing decisions, drive store traffic and ultimately sales of ASICS products.

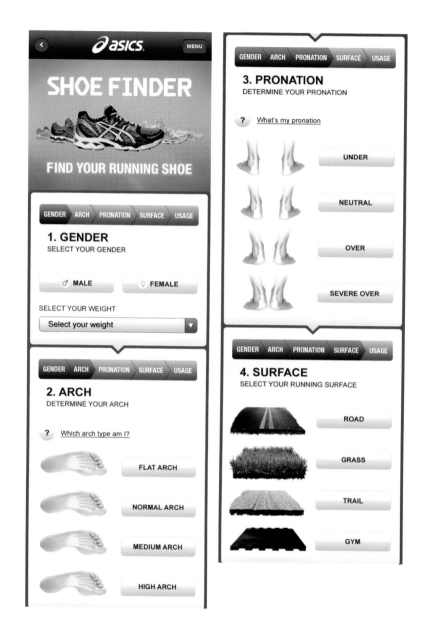

22 Percent mobile traffic triggering analytics goal

51 Percent traffic from iPhone

3 Minutes average user-time

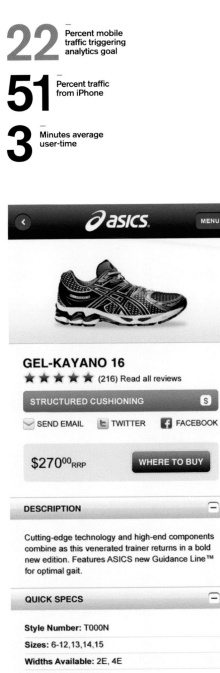

GEL-KAYANO 16

★ ★ ★ ★ ★ (216) Read all reviews

STRUCTURED CUSHIONING   S

✉ SEND EMAIL   🔲 TWITTER   📘 FACEBOOK

$270⁰⁰ RRP   WHERE TO BUY

DESCRIPTION ⊖

Cutting-edge technology and high-end components combine as this venerated trainer returns in a bold new edition. Features ASICS new Guidance Line™ for optimal gait.

QUICK SPECS ⊖

Style Number: T000N

Sizes: 6-12,13,14,15

Widths Available: 2E, 4E

2E Width: 8-14,15,16

4E Width: 8-14,15,16

Weight: 358 grams*
Weight is based on Mens size 7 shoe

The Solution

If we wanted users to take advantage of our mobile platform, it had to be simple, fast and provide bite-sized pieces of clearly defined information that would take users down one of many lead generation paths to find the right product to suit their needs, at the closest store stocking it.

So the site was designed specifically around simplifying the educational experience, with all content, navigation and creative focused on providing those simple, bite-sized pieces to be consumed on the move, with clear paths to finding a local store, matching shoes to your running requirements and reading customer reviews in-store.

The Results

The ASICS Mobile website launched to much fanfare, a category first in Australia, delivering a best practice mobile experience that harnessed almost 90 percent of the entire website's content, helping to deliver a seamless brand experience across multiple digital platforms.

With mobile traffic currently at 9 percent of all website traffic (and growing rapidly), ASICS are well positioned to take advantage of the huge mobile growth forecast for the next 12-24 months. Already the mobile site converts 22 percent of all visits into goals, with 51 percent of those being from iPhone users, spending three minutes per session on average.

Citi iAd

"The real success in Citibank being a part of the iAd launch was about them coming out strong post-bailout. Just one year prior, public perception was at an all time low, so participating in something beautifully designed, fun and cutting edge really placed the brand in a forward-thinking, leadership position."
Steve Farrell, Executive Creative Director,
Publicis Modem

The Brief

By signing on as one of the four launch partners for Apple's revolutionary iAd platform (along with brands like Nissan and Pepsi), Citibank was looking to extend their recently launched "What's Your Story?" campaign into the mobile space in a big and innovative way. They wanted to take the core of their new brand narrative, "What's your story? Citi can help you write it", and put it straight on to people's iPhones in a way that was beautiful, entertaining, actionable and immediate: the kind of superlative experience that people expect from their iPhone but not necessarily from Citi.

The Challenge

Later summer in 2010, Citibank was facing two major hurdles – parity competition and the aftermath of the TARP bailout. With rivals not only from other traditional banking institutions, like Chase and Bank of America, but also from direct credit companies, like Discover and American Express, it was important to set Citibank apart by highlighting the personal, relatable aspect of fulfilling customers' dreams and aspirations through the sharing of stories.

But with the extra scrutiny being applied to their every marketing move due to the TARP bailout, Citibank needed to be sure that the experience they were delivering was both brand boosting and business directed.

Client
Citi

Credits
MIR
www.mir.mx
Publicis
www.publicisusa.com

Awards
FWA Mobile

35 Percent banner
click through rate

7 times click through
rate vs online
banners

1 Minute "Story Sparker"
engagement time

The Solution

Create an iAd that pushed the possibilities of the platform by showcasing the many ways that Citibank can help people "write their own stories" either through inspiration or utility. By creating an immersive, explorable 3D street scene environment, we were able to create an iAd that felt rich and on-brand but still had the ease of an intuitive UI that people expect on an Apple device. In one click, users could watch other people's stories emotionally told, search for bits of inspiration nearby using geo-location or simply apply for the card that was right for them. By striking the right balance of emotion and utility, the Citibank iAd showed that a "service narrative" was not only possible, but absolutely necessary for brand building in the mobile space.

The Results

Beyond the fantastic word of mouth and press regarding the visual aspects and the user experience, we actually got some positive tangible results. The actual iAd Banner CTR, or Click Through Rate, for the Citi iAd was 35 percent, seven times the CTR for online banners. And as time moved on, this number did not diminish, indicating user interest was not a passing fancy related to the "newness" of the iAd concept.

The "Story Sparker" (The Street Scene which is the source of the greatest amount of rich content) was not surprisingly the focal point for engagement within the iAd. And in general, the average time for users to be engaged with the ad was greater than one minute, far surpassing other voluntary engagement media channels.

RedBull.com: Mobile Version

PICK OF THE DAY

Dec 03, 2010
Who is Levi LaVallee?
Who is Levi LaVallee? This New Year's Eve, Levi LaVallee, a seven-time Winter X ...

More Pics   >

THE LATEST STUFF

ARTICLES   PHOTOS   VIDEOS   EVENTS

Dec 06, 2010
Seb Named Racing Driver Of The Year
There were more wins for Red Bull Racing last night at the Autosport A...

More Articles   >

Client
Red Bull GmbH

Credits
Red Bull Media House GmbH
Cellular GmbH

Awards
FWA Mobile

"I was seriously impressed to see such a huge amount of content seamlessly working on mobile."
Rob Ford, FWA

The Brief

Red Bull, known for its Energy Drink available in more than 160 countries, is a true global brand and has a strong footprint in the digital world: whether through its content-rich website redbull.com, its social media activities on Twitter or Facebook, where it has millions of followers, or the mobile applications Red Bull has developed for iPhone and iPad or other smartphone platforms.

The natural step to increase the user experience for mobile devices was the creation of a state-of-the-art mobile experience for all content provided on www.redbull.com.

The Challenge

Red Bull's global websites are all about delivering high-quality content from the world of Red Bull. Knowing that Red Bull's audience is truly global, using all types of mobile devices with different sets of capabilities, the challenge was to reflect the premium content experience from the web on a wide range of mobile devices. Delivering an optimal video experience was a key requirement as well as covering all Red Bull local websites and language versions.

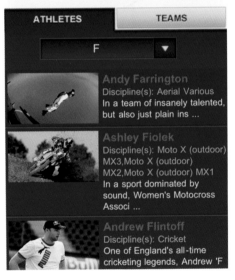

Red Bull TV

Lean back and be entertained with the best content the world of Red Bull - from action sports and motor sports to music and lifestyle. Get the Red Bull TV app for your iPhone to tune in!

 DOWNLOAD THE APP TO TUNE IN

WHATS ON TODAY

 **Daddy Yankee**
Daddy Yankee, is a Latin Grammy Award winning Puerto Rican reggae ...

Red Bull Surfing Project

The Solution
Cellular GmbH and Red Bull Media House GmbH have developed a global concept fitting all Red Bull mobile sites. Due to the different needs of modern smartphones versus regular mobile phones, two main versions have been created.

First, a touchscreen version was created using the latest mobile web technology to offer an optimal multimedia browsing experience. It allows the use of full-screen dimension, touch gestures, high-quality video encoding and full integration of Facebook's "Like" buttons.

The second version is optimised for non-touch mobile phones. While still offering all the video content available in appropriate formats, the site is optimised around text-based content and images. This ensures that older devices too can enjoy a satisfying presentation of the content.

The Results
Within a period of two months, the traffic from mobile devices to the mobile-optimised version (m.redbull.com) increased more than tenfold. The increase did not impinge on traffic from Red Bull's website but instead encouraged additional traffic by offering a great user experience for mobile users. Due to the tight integration with Facebook, an increasing amount of mobile traffic comes from social media.

1,000
Percent plus
traffic increase

0
Loss of traffic

Videos

 Dec 01, 2010
Fun ride in Buenos Aires
Duration: 2 min 31 sec

 Nov 30, 2010
Backyard Digger 2010 at Mote Park
Duration: 4 min 59 sec

 Nov 29, 2010
Capital Homecoming For Our World Champion
Duration: 2 min 14 sec

 Nov 26, 2010
Trendspy: Show 2
Duration: 6 min 17 sec

Nov 25, 2010
Meet Milton Martinez

Carissa Moore

D.O.B.:	1991-08-27 12:00:00
Location:	Honolulu, Hawaii, USA
Discipline(s):	Surfing

About Carissa

Depending on which language you're speaking, the name "Carissa" can mean "beloved," "grace," or "kind." But if you're speaking the language of surfing,

BabyCarrots.com
Mobile/Xtreme Xrunch Kart

"By the third time I heard the metalhead grunt that Baby! Carrots! are EXTREME!, I finally got how brilliantly self-referential it all was... The entire campaign is kind of genius. It's certain to attract attention and become a talking point."
Riddhi Shah, Salon

Client
Bolthouse Farms

Developer
unit9
www.unit9.com

Agency
Crispin Porter + Bogusky
www.cpbgroup.com

Awards
FWA Mobile; Creativity Online

The Brief
How do you re-brand baby carrots from a boring, old veggie into the ultimate snack-your-face-off crunchy munchy?

The Challenge
When you think "exciting snack," baby carrots don't come to mind. Like most veggies, baby carrots are forsaken to the refrigerator crisper, never to be seen or heard from again. Of course, junk food doesn't have this problem.

Their advertising is just as extreme and over-the-top as their marketing budgets. So how do we get baby carrots out of the bottom drawer and make them a sexy, top-of-mind snack alongside the likes of Cheetos and Doritos?

16.38 — Minutes
average time

33,000 — Downloads

75 — Countries
playing

500 — Million PR
impressions

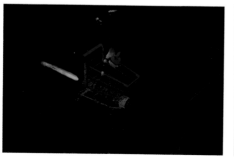

The Solution

BabyCarrots.com Mobile serves up all the junky good content built right into a giant baby carrot. Side scroll across it to check out tasty morsels like baby carrots' new junk food packaging and commercials, the exclusive YouTube series Munchies: The Late Shift with Chip & Abdul and the game trailer and link to download Xtreme Xrunch Kart.

XXK is the world's first-ever mobile game powered by crunching baby carrots directly into your iPhone mic. Using a crunch-detection algorithm dubbed "Crunchonics Technology," gamers crunch real baby carrots into their iPhone or iPod Touch mic to control a rickety shopping cart strapped to a rocket engine.

Every real-life carrot crunch translates into gnarly speed boosts, radical air and badical tricks, allowing your animated "xtreme hero dude" to clear massive pits, skip across rooftops and soar over buildings with the occasional pterodactyl flying overhead. Most importantly, there's a buttload of explosions.

The Results

Despite only officially being promoted in two small test markets to date, people around the globe couldn't get enough of baby carrots on their mobile devices. In fact, those surfing the webkit version of BabyCarrots.com spent an average of 16.38 minutes gobbling up content.

On top of that, Xtreme Xrunch Kart was downloaded nearly 13,000 times in the first week, and has been downloaded and played in more than 75 different countries worldwide. Both helped the Eat 'Em Like Junk Food campaign garner more than 500 million PR impressions around the globe thus far, including coverage by the Associated Press, USA Today, CNN, Washington Post and New York Times.

Nissan LEAF iAd

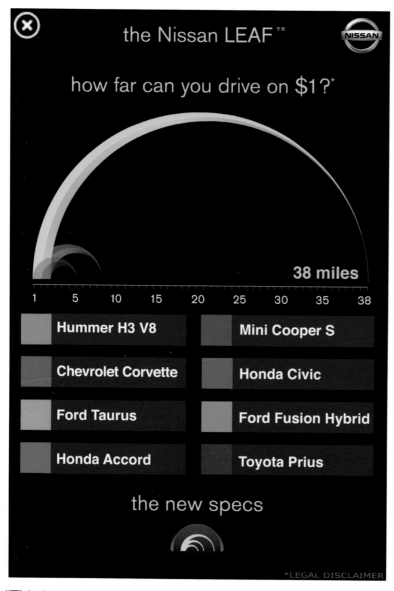

"This is a pretty doggone compelling way to get your point across. And it's fun."
Steve Jobs, Apple

Client
Nissan North America

Credits
TBWA\CHIAT\DAY
www.tbwachiat.com
OMD
www.omd.com

Awards
FWA Mobile

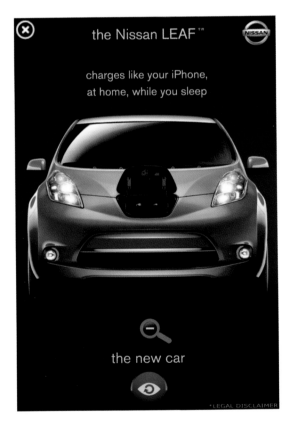

The Brief

Gas engines have ruled the road for the past 100 years, and we've all experienced the benefits. But increasingly, we've also seen the downside. Smog, high gas prices, oil spills, even political fallout from our reliance on foreign oil. The only real alternatives have been hybrids, and that's not much of a change at all.

Until the 100 percent electric Nissan LEAF. It's literally the world's first mass-produced electric car, priced for the masses. It uses no gas, has no tailpipe, and thus produces zero tailpipe emissions. It plugs in to charge overnight like a cell phone, and even talks to your phone. A vehicle as revolutionary as this deserves a revolutionary way to advertise it.

The Challenge

In April 2010, Steve Jobs announced the new mobile advertising platform, iAds. Designed specifically for the iPhone, iAds are premium brand experiences that live within existing iPhone apps. They are inherently unobtrusive, but once a user opts in, fully immersive, employing most of the phone's capabilities.

Which all sounds great, but at the time no one knew what an iAd was. Yet it was immediately apparent that enthusiastic iPhone owners who were excited about iAds were a perfect fit for the Nissan LEAF – tech savvy and cutting edge, otherwise known as early adopters. We just had to figure out how to create an iAd, since no one ever had before.

The Solution

The key to making the iAd work was not so much in solving the technical challenges, but in crafting a compelling brand narrative that flowed seamlessly into every corner of the iAd.

Our Nissan LEAF print campaign describes it as "the new car". We took the idea of new as inspiration to redefine everything related to the vehicle – from its specs to its fuel efficiency rating. We offered a chance to win a car through the iAd. And we developed a future history of the Nissan LEAF as an epic video to set the whole thing up.

All of this messaging was packed into six easy-to-digest sections, and delivered via a carousel navigation that served the dual purpose of showing off a 360 view of the car.

The Results

Our first indication that we'd hit the mark was when Steve Jobs chose to demonstrate how iAds work by showcasing the Nissan LEAF iAd at the Worldwide Developers Conference in June 2010. The volume of press and social media conversation from that event was overwhelming.

When the iAd started appearing on iPhones, the response was even better. Customers spent an average of 90 seconds with the ad – ten times longer than comparable online ads. And people chose to "tap" on the LEAF iAd five times more frequently than regular online display ads for the LEAF.

The Nissan LEAF iAd has become the gold standard for iAds, and a case study for other brands in how to achieve success in this new medium. Best of all, the Nissan LEAF sold out its first year of production in just a few months.

5 Times more "tap throughs" than previous LEAF online ads

90 Seconds average time on ad

10 Times average time more than other digital ads

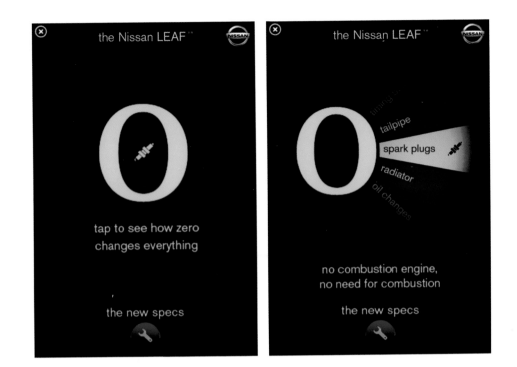

Ruby Tuesday Mobile

Client
Ruby Tuesday

Credits
AgencyNet
www.agencynet.com

Awards
FWA Mobile

"Mobile is a no-brainer for casual dining. The customer is out, they're hungry, they're interested in our restaurant… the mobile web lets us make it as easy as possible for them to sit down at our table."
Brian Chiger, Digital Strategist, AgencyNet

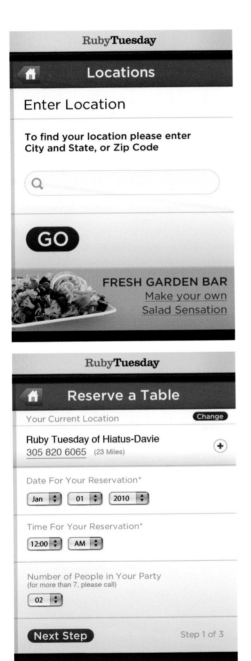

The Brief

Looking to distinguish itself among competitors in the casual dining restaurant space, Ruby Tuesday began a transformation in late 2007 to evolve the restaurant into an upscale, modern dining experience.

Founder and CEO Sandy Beall began by revamping the restaurant's décor and menu, and in partnership with AgencyNet, brought the story of the new Ruby Tuesday to life in the digital space. In May 2010, the brand launched the new RubyTuesday.com – built using a standard framework to allow it to function across all multiple device browsers (including tablet), the experience on a mobile's smaller screen was less than ideal. And while the new RubyTuesday.com would have been viewable on mobile devices, we knew we needed to provide a more customised, accessible, and streamlined experience for the user on the go.

The Challenge

Before choosing to visit a Ruby Tuesday, a prospective guest wants to know certain things, and increasingly consumers expect to be able to access that information from their mobile device. Our challenge was to streamline the user experience of RubyTuesday.com to serve the unique needs of the mobile user. What's on the menu? Where is the closest location? Will there be a table available when I get there? Are there any special offers I should know about?

In addition to designing an experience that answered these core questions, we also envisioned an experience where guests could access custom offers and activities in the restaurant environment delivered straight to their palm.

Ruby Tuesday

STEAK & LOBSTER
MAC N' CHEESE
Try it today!

Latest Offers >

Locations >

The Menu >

The Bar >

Reserve a Table >

So Connected >

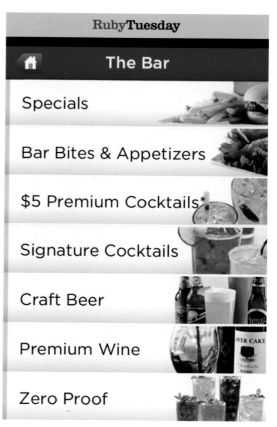

Ruby Tuesday

🏠 The Bar

Specials

Bar Bites & Appetizers

$5 Premium Cocktails*

Signature Cocktails

Craft Beer

Premium Wine

Zero Proof

The Solution

Our solution re-imagined the experience of RubyTuesday.com for the mobile user.

A geo-aware location finder lets users on-the-go find a nearby Ruby Tuesday with fewer than two clicks. Localized food and bar menus let users contemplate their order and salivate over their favorite items before ever setting foot in the door. We also gave users the ability to reserve a table directly from the mobile web experience and become a part of Ruby Tuesday's So Connected CRM program.

The finished experience is much lighter than loading the conventional RubyTuesday.com website, allowing users to load the page faster and quickly find any information they're looking for.

The Results

In the first three months since its launch, 100,000 customers have used our site to find directions to the nearest Ruby Tuesday. 70,000 of those users decided what to eat before they got there.

In response, fans have stood up and noticed the new, more sophisticated Ruby Tuesday. Since we began the revamp of their digital presence (including their mobile, .com and social presences), Ruby's social networking fan base has grown to 300 times its former size and, as of this writing, is reaching upwards of 300,000.

Most importantly, the mobile site has worked alongside digital and offline experience to make the last quarter in 2010 Ruby Tuesday's best sales quarter in four years. Now… Who's hungry?

100,000 Using site to find nearest Ruby Tuesday

70,000 Deciding what to eat before arriving

300 Percent social network increase

300,000 Social network size

Visa Signature iPad Wallet

"More than any iAd I have seen... the 'This is No Ordinary Wallet' creative is in harmony with the device's touch interface and nominally magical toy-like allure."
Steve Smith, Mobile Insider

The Brief
Use iPad functionality to bring the Visa Signature perks to life while allowing the user to "play" with their iPad.

The Challenge
Visa Signature is a more rewarding way to pay than other rewards cards because it enhances your everyday with relevant perks you don't have to earn. But many Signature cards holders don't know their card is a Signature card. By featuring the perks, we are trying to entice people to look into their wallets – and therefore raise awareness.

Client
Visa

Credits
AKQA
www.akqa.com

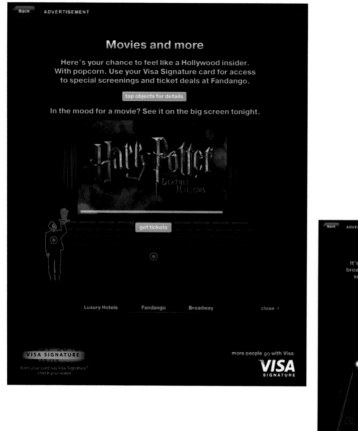

500,000
Plus impressions generated

8,000
Plus engagement touch points

The Solution

The solution is the iPad "Wallet", where users can interact and see what experiences and perks they have just by being a Visa Signature cardholder. The three perks we focused on were with partners Fandango, Broadway.com and the Kiwi Hotel Collection. This came to life by taking a twist on the existing templates available to include pop-up animation in HTML5, dual navigation, robust site architecture, hotspots and videos and third-party APIs.

There are technically multiple ways to explore more about the perks (i.e. watch the latest movie trailer on Fandango, search hotels and availability on Kiwi) but the main call to action drove consumers to redemption.

The Results

The Visa Signature iPad ad received critical praise including being named one of the top three coolest iPad ads in the market by the *Wall Street Journal*. Additionally, the Visa Signature iPad ad was written up in industry publications such as *Creativity Magazine* and *MediaPost*. The latter publication has been quoted as saying "More than any iAd I have seen… the "This is No Ordinary Wallet" creative is in harmony with the device's touch interface and nominally magical toy-like allure" and that "this ad… succeeded in communicating product attributes and a bit of brand character. This ad illustrates the range and depth of Visa Signature services in ways that mere video or print alone couldn't."

Auto Trader iPhone App

"We believe users will find the link between 'real world' and 'digital world' very compelling. Using the iPhone users can now search for real-time digital information relating to a car sitting in front of them, a step which combined with the depth of cars listed on Auto Trader takes the car-buying process to a whole new level."
Nick Gee, Director of Mobile at Auto Trader

The Brief

In all our work with Trader Media, we have tried to use the concept of a "delighter" – a cheeky surprise that causes people to smile, without distracting from the main message. The idea is to pepper Trader Media's digital presence with "delighters", making it stand out from the competition and building a rapport with their ten million-strong customer base.

The brief here was to create an application that delivers a unique experience for Auto Trader iPhone users, building a brand experience in a new marketing channel. Use the unique features of the iPhone technology to create an experience that surprises and delights the user. We needed to take the potentially pedestrian experience of searching for a second-hand car and make it fun, engaging and fresh.

The Challenge

The world of motoring has moved on massively since the first cars took to the roads. However, the way people buy cars has stayed the same since the 19th century.

It was time for a change. We wanted to give people the chance to buy when they were actually thinking about cars – when they see one on the street or hear the hum of an engine.

Client
Auto Trader –
Trader Media Group

Credits
SapientNitro, London
www.sapient.com/
en-us/sapientnitro.html

Awards
FWA Mobile; Campaign BIG Awards;
Cannes Lions; London International
Awards; OMMA Awards; W3 Awards

The Solution

The strategy was simple: rethink the user journey for finding a second-hand car to include the mobile channel. Use the opportunity provided by mobile technology to change how people interact with the brand.

So we built an iPhone app for Auto Trader, the UK market leader for buying and selling used cars. The app allows people to take a picture of any car and instantly get information about that make and model. They can read reviews, view ads, compare prices, see where cars are located, get directions and call up to arrange a test drive. All from just a photo.

The Results

We managed to repackage Trader Media's core business function of searching for second-hand cars into a fresh, new experience. Consumers engage with the brand in a very immediate way, linking the real world in front of them with the online world of motoring. The app puts Trader Media at the heart of a revolution in the way we think about cars.

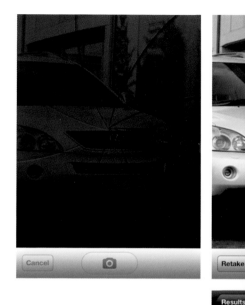

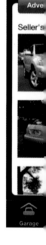

#1 UK Lifestyle section of the App Store after one day

#1 UK App Store after three days

20 Press articles

233 Revisions

10 Go-live meetings

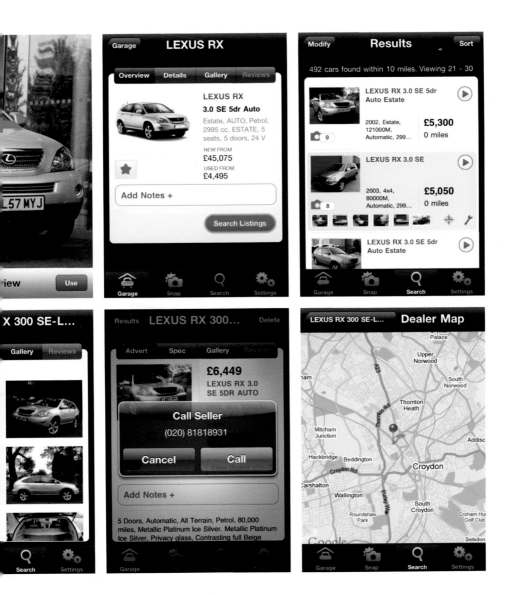

Nigella Lawson
Quick Collection

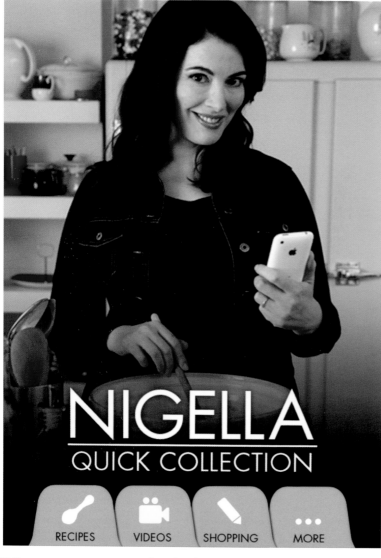

"Like us, you were perfectionists striving for the best possible design, UE, accuracy and experience. With voice control, a feature something we would have never considered without your input, you even added a healthy portion of AKQA technology sparkle."
Jonathan Davis, Random House

The Brief

Nigella Lawson wanted to bring speedy, delicious recipes to the mobile space in an app that was both practical and beautiful for the user. AKQA was tasked with creating an app that could take its place in the heart of the kitchen, using the capabilities of mobile to make enjoying Nigella's recipes a simple, stress-free social experience. The target audience for the application was iPhone users with an interest in cooking a range of delicious recipes.

The Challenge

We wanted this to be more than just another cooking app, and instead to raise the bar in the Lifestyle category. Not only did the application need to be aesthetically stunning, but it also needed to be practical. The content speaks for itself; therefore creating a flawless user experience which ensured using the app on a regular basis would be nothing short of a delight was paramount. And to top it all off, the application had a very tight timescale for delivery, meaning production must be quick as well as perfect.

Client
Random House

Credits
AKQA
www.akqa.com

Awards
BIMA 2010

70 Unique Nigella Lawson recipes

1 World's first voice-activated cooking app

#1 Lifestyle app in UK, NZ and Australia

The Solution

Nigella Quick Collection is an easy-to-use, sumptuously designed mobile app that helps the user create speedy, delicious meals while keeping the phone free of sticky fingerprints. Over 70 of the domestic goddess' mouth-watering recipes are available, ready to be searched by ingredient, mood or cookbook.

Innovation was at the core of the design. Users avoid sticky situations as they keep their phones pristine while cooking, with easy voice navigation through the recipe steps – a world first for a cooking app.

There's the opportunity to personalise recipes with voice or text annotations, tell friends what's cooking on Facebook and invite them to join you for dinner – all direct from the app.

There's also an inspirational video from Nigella, handy "how to" guides and a feast of photography, to keep you drooling all the way to the dinner table.

The Results

The app received huge press acclaim, with the *Times* describing it as "sleek, elegant and sophisticated" and adding "What did you expect?" This app is recognised as a benchmark within the Lifestyle category, with many other cooking apps "borrowing" its innovative features: it has therefore helped to define the category.

Nigella Quick Collection became the number one Lifestyle app in the UK, New Zealand and Australia and has consistently been in the top 25 Lifestyle apps during 2010. As a publisher, Random House will not disclose commercial results, but has stated many times to the trade that this is its most successful iPhone application ever.

McDonald's France
iPhone App

Client
McDonald's France

Credits
Mobile Dream Studio
www.mobiledreamstudio.com

"Very well done. The list of cinemas near each restaurant... What a great idea!"
Claisanc, App Store user

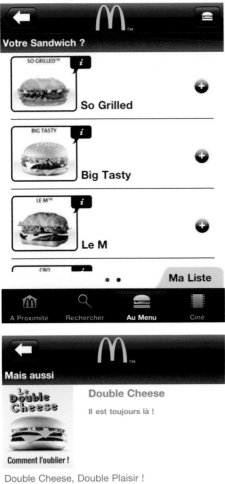

The Brief

With the overwhelming development of the iPhone phenomenon in France as in the whole world, it was unimaginable for the fast-food industry leader not to play in this field. As a consequence, McDonald's France asked Mobile Dream Studio to develop an iPhone application that would be an extension of their digital strategy. The aim was to offer a mobile experience to their consumers, who had the will to find the brand restaurants easily and to have information about their products. Integrating such an innovative dimension would strengthen the relationship between the brand and its consumers: by offering them a way to have McDonald's France permanently at their fingertips.

The Challenge

Being a leader is not so easy and there was a lot at stake. McDonald's France and Mobile Dream Studio had to live up to consumers' expectations to prove that this application would not be one amongst the others. It was necessary to propose a free experience to users and to bring them a real benefit. In that way, the brand wanted to offer to its consumers an easy and complete application: it had to be functional, practical and emotional. To fulfil this goal, the content of the application was determined according to the consumers' social patterns of behaviour, without losing track of what they were expecting.

The Solution

The core of the McDonald's France application lies in the organisation of users' visits to McDonald's restaurants. They can find these by choosing between a geo-localisation function and a search function. Then they receive the restaurant information, such as the services it offers, its address and telephone number, the current track played (for the restaurants equipped with Deezer) and directions to find it.

As an extension of this, an entertainment dimension enables users to receive a cinema editorial selection, to look up the movie's release, to find cinemas and to consult their schedule.

In addition to these functions, users can consult a generic list of the products and get information about them. On top of that, a notepad system helps users to create their own shopping list to prepare their visit to restaurants, a list which can be sent by email.

The Results

The application was a real success. It was launched in September 2009 and has been downloaded almost 2,000,000 times since this date, even though it is only available in France for the moment.

This enthusiasm provoked in many consumers and iPhone users led it to be ranked number one as the most downloaded application for a brick and mortar brand in France. The application has been in the top ten since its release in the Lifestyle section of the French App Store.

Furthermore, coupled with this download data, the frequency and duration figures indicate the way the application is used. McDonald's France and Mobile Dream Studio succeeded in developing an application useful and necessary enough to be used regularly, and, in 70 percent of cases, for more than three minutes.

3 Minutes on site for 70% of users

800,000 Sessions per month

1,921,758 Downloads

Domino's Pizza
iPhone App

"The launch of our iPhone app was in line with our strategy of being a digital pioneer and putting a bit of Pepperoni Passion into more people's lives. We are genuinely delighted with the results and the impact on our business."
James Millett, Multimedia Manager, Domino's Pizza

Client
Domino's Pizza
UK & Ireland

Credits
Mobext
www.mobext.com

The Brief

Domino's Pizza is a youthful, fun brand, which uses innovative digital media to engage with its customers. Domino's was the first pizza delivery company to launch interactive and Internet ordering in 1999 and introduced pizza tracker, which enables customers to track their pizza from ordering right through to reaching their front door, in 2009.

This history of innovation and the huge growth in online ordering over the past few years meant an iPhone app was the obvious new addition to Domino's digital armoury. However, it had to have the right features and functionality to make sure it continued to provide the innovation and engagement Domino's customers are accustomed to.

The Challenge

The challenge was to create an app which made it easy and fast to order a pizza, but at the same time offered something more to engage users and reinforce the personality and key attributes of the Domino's brand. Our website is already optimised for mobiles but we wanted to make it even quicker and easier for customers to get a piping hot pizza, wherever they are.

While the main function was ordering, it was essential the app also included other features to maintain the brand's funky, playful characteristics. In true Domino's style, the new app does just that and also brings a bit of fun, which builds on the great success of our interactive pizza tracker.

The Solution

The app uses GPS to automatically locate the nearest Domino's store. Users can then scroll through a carousel-style menu to select their favourite pizzas, sides, desserts, subs and drinks. Pizzas can be customised by pinching to resize the base and scattering on tasty toppings before sending the order straight to the store. iPhone users can pop their pizza in the oven and swipe or blow away the steam for a chance to win money off their next order. Then, while the pizza is in the real oven, users can keep tabs on its progress with the real-time pizza tracker.

With around 75 million pizza combinations on a typical Domino's menu, the app also provides a great way to "Create Your Own" with its pizza slot machine feature. Simply shake your phone and pick a totally random combination of base, sauce and toppings. The choices can be spun again or locked and sent as an order straight to the store.

The Results

The app is everything we'd wanted it to be. Customers can order from Domino's easily and quickly on the move and have a bit of fun with the oven game and pizza slot machine functions. It sums up our brand perfectly. The number of downloads the app has received to date has been phenomenal and we are very pleased with the orders it has generated so far.

On the back of demand for this app, we're also now launching an android and iPad version to enable everyone to have a "Domino's store in their pocket"!

1 Million plus GBP revenue generated in pizza sales

400,000 Downloads

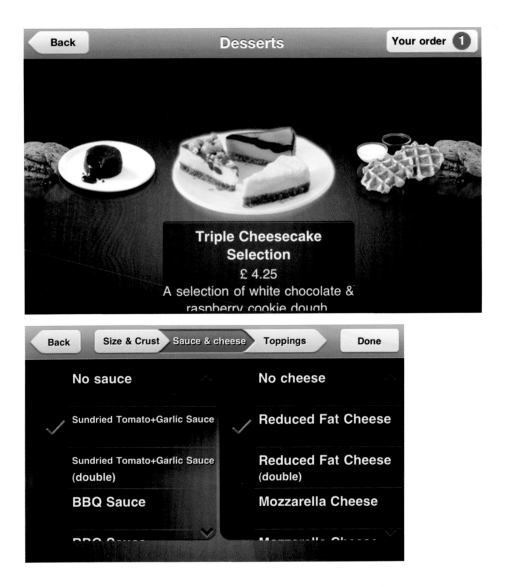

Nissan Juke iAd

"You may have already seen this, but dead set, if this isn't the best iAd campaign to date, then it's certainly the best ever iAd Demo video, and that might just be pretty dam tough to beat!"
Aden Hepburn, Digital Buzz Blog

The Brief

Our target: urban males in their mid-20s. For the first time in life they're their own masters, free from the confines of school or parental oversight. These guys want to take control. They're driven to get ahead, to make a name for themselves. To stand out on their own terms.

Our vehicle: the Nissan Juke. Something the designers coined "totally new". With the style, performance and agility of a lightweight sports car together with the robust stance of a CUV. An aggressive, polarizing exterior that's completely unique. And an interior with adaptive technology and a driver-biased cockpit. A car at the top of its food chain because it redefines what a car its size is capable of.

Take the Juke's mantra of innovation unleashed, and unleash it with advertising.

The Challenge

The LEAF iAd blazed the trail for what iAds could be. It was one of the first. And arguably one of the best. And it was the one Steve Jobs used to show the world what an iAd was.

So we had to follow that up. And with a car that no one had heard of, and some even called more of a "joke" than a "Juke". A male millennial target with the attention span of a mayfly. A new medium on a new platform. Concepts that were technologically impossible.

Maybe making the 27th iAd ever isn't as hard as making the first one, but it's still a walk in the dark. But we needed to make it great – and maybe, even better than any before it.

Client
Nissan North America

Credits
TBWA\CHIAT\DAY
www.tbwachiat.com

Awards
FWA Mobile

The Solution

If we needed to unleash innovation for the Juke, what better way than to use the innovations of the iPhone to unleash the innovations of the Juke.

Users learned about all the Juke's cool new features, because that's what makes the clients happy, but ingrained them organically into a ridiculously fun and cool-looking ride, because that's how this target likes to learn.

We turned a boring task – getting donuts for a status meeting – into a super-cool action adventure in a motion-comic style mission to get donuts.

Turbo-charged getaways. Maxim hometown hotties. Breaking through glass windows. Bonus donuts with hidden exclusive content, all powered by iPhone actions like swiping, shaking and tapping. Plus a sweepstake for a free Juke, available only to those who complete the mission.

The Results

We built the first-ever comic book iAd. The first with a continuous storyline. And the first with a storyline users control.

It's what a creative director at Apple called "being able to taste chocolate" without eating any.

And while the Nissan Juke iAd is still new, according to our recent Twitter searches, no one is calling the Juke a "joke" any more. Instead, it's: "tap on my face" and "cool shoes" and "I need a Nissan Juke… I said NEED and not WANT". And changing the mind of a 22-year-old guy (who would rather listen to his mother than an ad, and never listens to his mother) may be a better success than any sort of click-through conversion.

1 — First ever Comic Book iAd

42 — Percent increase in tap-throughs over LEAF iAd

161 — Photoshop files

Waitrose

"We focused on our stengths of excellent food content and expertise and combined that with the key trend in mobile to share content and information via email and social networks. This enabled us to create a mobile experience that engages with our existing customers and, at the same time, reach out to new customers who may never have had an interaction with the Waitrose brand before!"
Fiona Hall, Manager, Innovation, Waitrose

Client
Waitrose Ltd

Credits
YOC UK
www.yoc.com

Awards
iTunes UK app of the week (Aug 2010); featured by Apple as one of their top apps of 2010 in Cooking

The Brief

Waitrose has been established as one of the UK's top high-end food retailers for decades. With a reputation for impeccable customer service and unparalleled quality Waitrose sought greater penetration from their brand image and a unique and different approach to mobile from their competitors. Where better to achieve this than through mobile Internet and applications.

The Challenge

As an established high-end brand, Waitrose's profile may not seem immediately compatible with the cutting-edge mobile market. We needed to make the brand's presence both engaging and sticky by producing a mobile experience that would become indispensable to the user. The obvious place to engage users was through mobile commerce, a solution unfortunately encumbered with back-end complications. As such we moved to interact with the consumer where their relationship with food is at its most intimate: in the kitchen. The second aspect was driving users into Waitrose stores and engaging with Waitrose products.

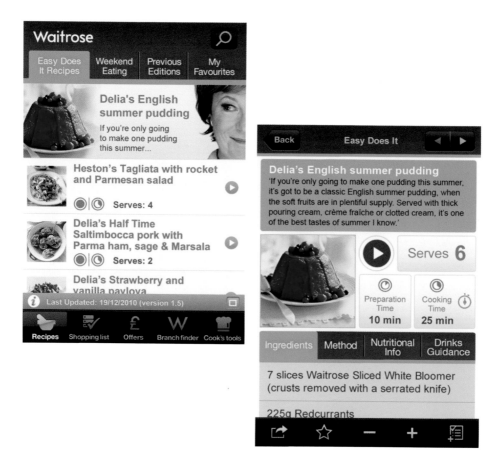

308,878 Downloads

1 Download every four seconds in first weeks of launch

The Solution

Drawing on Waitrose's long-established recipe library and its relationship with renowned celebrity chefs Heston Blumenthal and Delia Smith we developed a kitchen utility app. The application delivers a wealth of recipes which are updated weekly and include contributions from the aforementioned celebrity cooks. All the recipes are shareable via email and social networks so the Waitrose brand and content had reach beyond the app itself.

We were able to deliver a richer experience through the inclusion of kitchen utilities including measurement converters and cooking timers, designed to make the app an indispensable kitchen utensil as well as a recipe library. The inclusion of instructional videos for everything from filleting a fish to bringing a pan to a rolling boil took this further.

To drive users into stores we included a store finder with branch information and opening hours. We also included current special offers and a digital service counter that aimed to emphasise the quality of Waitrose's produce. As well as a shopping list tool, for customers to use whilst in store to shop for offers and recipe ingredients inspired by the app.

The Results

The application proved to be a runaway success achieving more than 300,000 downloads to date. It was selected to be iTunes UK's app of the week. The project undoubtedly delivered Waitrose into the mobile market in emphatic style. One user even reported that he had lived on nothing but the app's recipes for an entire week! The level of user engagement is undeniable.

Gordon Ramsay
Cook With Me

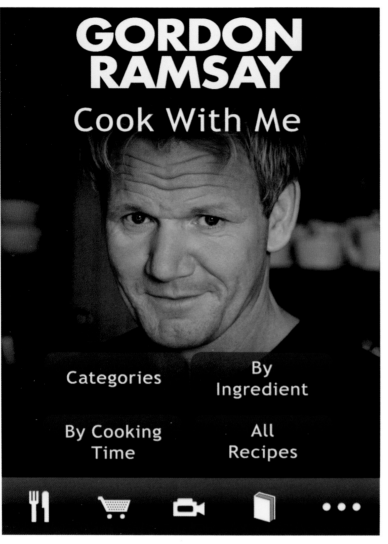

"A huge amount of work went in to Cook With Me and I'm very proud of the results. It's a cooking app to swear by, not at."
Gordon Ramsay

The Brief

The brief for Gordon Ramsay Cook With Me – the superchef's first ever cooking application – was to create an easy-to-use cooking app that would enable foodies all over the globe to cook with the world's greatest chef. Better than any recipe book – because you get video as well as words and pictures – using the app should be like having Gordon Ramsay at your side every step of the way, sharing his chef's secrets and showing you just how easy it is to cook delicious meals using a set of tried and tested recipes that will never let you down.

The Challenge

Gordon's app wasn't the first in the marketplace, so it had to be the best. It also had to capture Gordon's trademark wit and attitude, and inspire and help users to cook Gordon's recipes, from the simple sort of dishes Gordon cooks at home for his friends and family, to the sophisticated recipes on offer at his many award-winning restaurants. As well as including all the features users would expect from a top-end cooking app – an interactive shopping list, wine pairing suggestions, a broad range of recipes – the app had to feel fresh and exciting, chiming with Gordon's reputation as a world-class chef.

Client
One Potato Two Potato

Credits
Created/Developed
One Potato Two Potato Ltd
www.onepotatotwopotato.tv
M&C Saatchi Mobile
www.mcsaatchimobile.com

Awards
Top 10 Lifestyle App

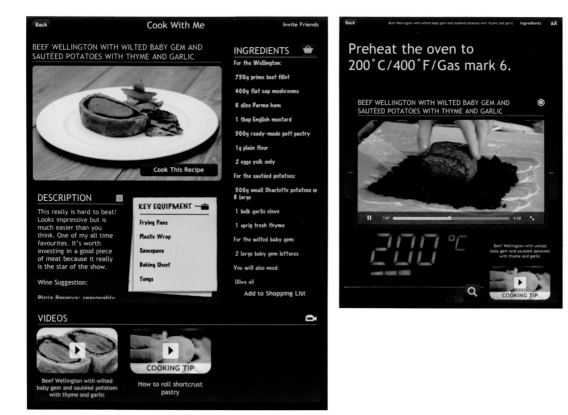

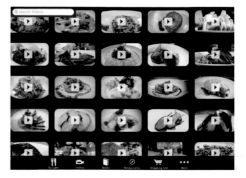

10
Top 10 iPad
Lifestyle app since
launch

30
Countries worldwide

56
How-to video
recipes

The Solution

Unlike the vast majority of other cooking apps on the market every single recipe in the app has a beautifully shot video to watch for inspiration, as well as a set of great video cooking tips. Users can pick a dish to cook from the 56 recipes on offer in a variety of ways: from one of five recipe categories, by the amount of time they have to cook, by difficulty, by the key ingredient they want to cook with, or by season. iPod or iPhone users can simply shake their device and be offered a suggestion for a great seasonal recipe, and when they finish cooking, can use the device's camera to compare their dish with Gordon's, then share the photo with their friends on Twitter or Facebook.

The Results

Gordon Ramsay Cook With Me for the iPod is widely considered to be the best cooking app on the market, and the iPad version takes things to a whole new level, with HD video and an enhanced user interface that makes brilliant use of the iPad's larger screen size. From cooking for one to a full-scale dinner party, Gordon Ramsay Cook With Me is the ultimate foodie cooking application. With delicious video recipe content and brilliant advice every step of the way, it's the only cooking app you'll ever need.

Promotional

**Introduction by
Ajaz Ahmed, AKQA**

Promotional Promotional

03

Promotional Promotional Promoti
Promotional Promotional Promoti
Promotional Promotional Promoti
Promotional Promotional Promoti
Promotional Promotional Promoti
Promotional Promotional Promoti
Promotional Promotional Promoti
Promotional Promotional Promoti
Promotional Promotional Promoti
Promotional Promotional Promoti
Promotional Promotional Promoti
Promotional Promotional Promoti
Promotional Promotional Promoti
Promotional Promotional Promoti
Promotional Promotional Promoti
Promotional Promotional Promoti

Promotional Promotional Promotional Promotional Promotional Prom
Promotional Promotional Promotional Promotional Promotional Prom
Promotional Promotional Promotional Promotional Promotional Prom
Promotional Promotional Promotional Promotional Promotional Prom
Promotional Promotional Promotional Promotional Promotional Prom
Promotional Promotional Promotional Promotional Promotional Prom
Promotional Promotional Promotional Promotional Promotional Prom
Promotional Promotional Promotional Promotional Promotional Prom

The next creative revolution

"You have part of my attention, you have the minimum amount" is one of the most memorable lines from the film *The Social Network*. Representing the voice of the next generation, it could also be the best way to articulate how most audiences react to conventional advertising of any kind.

"With a multitude of distractions easily available from just about any device that's connected, it's almost ridiculous to expect that audiences will automatically pay attention to advertising given the other options available."

There have been some amazing changes recently and one of the most profound is that it's never been easier for someone to scan past a message. With a multitude of distractions easily available from just about any device that's connected, it's almost ridiculous to expect that audiences will automatically pay attention to advertising given the other options available.

We now have audiences that have been raised with so much media in their lives that they know instinctively how to filter and manage it all pretty well. At the same time the media itself is becoming more intelligent at curating content and services of interest. The result is that audiences today are far more discretionary as they have the ability to dramatically edit what media they spend time with. Increasingly and all over the globe, the device they have most interaction with is one that is mobile.

Recent advances have given birth to extraordinary mobile devices with exceptional power. Over time, further advances in materials sciences will create new kinds of form factor which will fuel the imagination for the development of better experiences. To create more desirable products, the best device-makers will have a preoccupation with experimenting with and exploring new materials. This obsession with hardware innovation will inspire new kinds of software ideas that take advantage of the spectacular capabilities, connectivity and mobility of the new devices.

The challenge for teams working to create inspirational work is to ensure that their philosophical approach should be about the audience wanting to applaud it rather than seeking out or stumbling across something more interesting. The best ideas have more respect for what the audience gets out of the work – how it will inspire, satisfy or motivate them – rather than simply bombarding people with meaningless messages, irrelevant functionality and clutter.

Great brands have always been about smart, artful storytelling.
It's just that today the canvas to convey a message has more
variations. Mobile has enabled storytelling to evolve so it's
interactive, non-linear and deeper, taking it to a new level, making it
more personal and often with unexpected and mind-blowing results.

The ideas that stand out are produced by pioneers who
say it like it's never been said before. These ideas are not slaves to
the rules. They are not about retrofitting old formats into a new
medium. The work is not approached with preconceived ideas and
sentiments, but rather a mindset of reinvention.

The best ideas jump from technology into the realm of
imagination. They are about engagement, entertainment and
creating indispensable utility. There's always an element of daring
combined with a true and sincere passion for each project.

**"Mobile has enabled
storytelling to evolve so it's
interactive, non-linear and
deeper, taking it to a new level,
making it more personal and
often with unexpected and
mind-blowing results."**

This chapter celebrates
work that leaps out, commands an
audience's attention and rewards them.
This is work that makes people react
emotionally, stimulates their curiosity
and has a wonderful simplicity and
visual impact to it. It is work that strikes
a chord so that people respond to it.

The best ideas make use of
the unique functionality, input and
interaction characteristics of the new
devices intelligently. Instead of the conventional interface paradigms
used on other platforms, these ideas offer more natural input such
as tap, swipe, spin, rotate and shake, producing experiences that
are intuitive, attractive and as forward thinking as the teams that
created them.

Shaping an attitude and moving people towards action
has always had far more to do with emotion than function. It is
about being believable and persuasive and expressing ideas in the
most meaningful way. Clutter, ugliness and bad design are a form
of visual pollution wherever they occur and for brands are life-
threatening. The work that leads is defined by its aesthetic and
experiential impact. However modern the device, the timeless
principles of beauty, honesty, simplicity and accessibility combine
to bring about a positive, enjoyable reaction when done well.

03 Promotional

Accessibility probably isn't the first word that pops into the head of a design team when working on a new brief and yet, in just about every category, it is the most accessible brands that are the most successful. This is especially true for work on mobile devices where human-centred design is vital.

"A constant flow of new people coming in with ideas that are rich with elegance, artful, interesting, rewarding and enjoyable, and will help to keep the work at the forefront of culture."

Mobile gives brands the opportunity to have a more intimate, tactile connection with the audience because devices provide the connection of touch and automatically feel more real, more personal and more familiar. There are many people who feel uncomfortable with a PC but are much more at home when using a mobile device.

The challenge for a creative team working in this environment is to always be working on something new while considering the interplay and collaboration between various disciplines. There should be a process of rigorous examination to enhance and perfect the product and the experience.

To open the eyes and ears nothing stirs people more than that which is new and fresh and unlike anything they have ever seen before. The next creative revolution has already started and it began on mobile. A constant flow of new people coming in with ideas that are rich with elegance, artful, interesting, rewarding and enjoyable, and will help to keep the work at the forefront of culture.

Ajaz Ahmed
AKQA

Bio.
Ajaz Ahmed
AKQA

Ajaz started AKQA with the founding values of innovation, service, quality and thought. AKQA is now the world's largest independent agency and the most awarded. It has been named Agency of the Year over 20 times and employs around 1,000 people with offices in the USA, Europe and China.

AKQA has created some of the world's most influential and iconic digital experiences. Ajaz is also executive producer of Jamie Oliver's 20 Minute Meals app for iPhone.
–
http://akqa.com

"The next creative revolution has already started and it began on mobile."

Nike True City

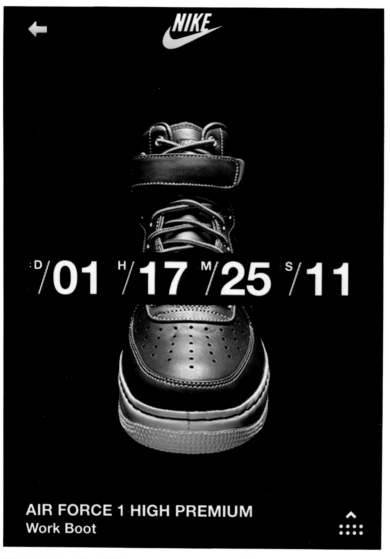

NIKE

:D /01 H /17 M /25 S /11

AIR FORCE 1 HIGH PREMIUM
Work Boot

"This is the first iPhone app to ever win an
FWA Site of the Day. This is a pioneering
app and an historical day for FWA."
Rob Ford, FWA

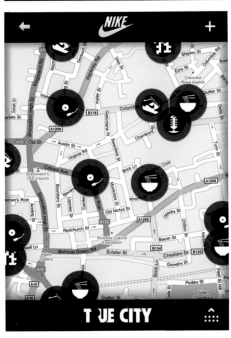

The Brief

In 1989, legendary Nike designer Tinker Hatfield created the original Air Max 1. Inspired by the unique "inside out" architecture of the Pompidou Centre in Paris, Tinker decided to cut a hole in the shoe's sole and reveal the signature Nike air capsule. 20 years later, in the summer of 2009, Nike launched the iAM1 campaign in support of the re-release of the classic and updated AM1 shoes.

In the spirit of the original AM1 design inspiration, Nike asked AKQA to devise a digital enablement piece that would allow fans to reveal their cities by sharing their own extraordinary journeys for others to discover.

The Challenge

The original pitch focused on one primary mantra: "Through the eyes of revolutionary people, we will make the hidden visible." We asserted that every city has its secrets, but few people have the key to unlock them – unlock the places, the art and technology, the living, beating heart of the real city. The Nike True City app, with its unique combination of "Insider" content, community contributions and advance access to the most exclusive Nike products more than opens the door to the city; it lets users dive in with all five senses.

03 Promotional

Client
Nike

Credits
AKQA
www.akqa.com

Awards
FWA; Cannes Cyber Lions;
IAB Creative Showcase;
Eurobest Mobile Grand Prix

Arches 477/478, Batemans Row, Shoreditch, EC2A 3HH

Known for his sometimes provocative and women-centric themes to his work, the Nike Sportswear x INSA "Looking for Love" exhibition was arguably centerpieced by his work on the "INSA Vortex" which engulfed the room.

Smaller works included some pieces which play on some of INSA's popular icons such as the usage of pink and element of femininity. In addition, some pieces INSA were removed from their original spots and also brought in for showcasing.

The Solution
Rather than a guidebook for visitors, the app delivers a powerful combination of premium, geo-tagged content. Users download the app and get an alternative taste of their city – updated in real time by real people – and become inspired to start tagging and broadcasting their own view of the city on the fly, using the latest iPhone geo-tagging and Facebook Connect technology.

It's a powerful two-way partnership between the brand and its audience. Inside the app, the BUZZ filter counts down to the coolest Nike product drops and events, and a hidden world of True City content can be hunted down and unlocked using the app's QR code reader.

The Results
Nike True City continues to inspire influencers and Nike lovers to share their true experiences of living in some of the world's most creative cities with the Nike True City community. The app has received thousands of contributions from passionate users and over a year on it continues to receive submissions from users with an insider's view of sport, life and culture across six of Europe's most innovative cities: London, Berlin, Paris, Milan, Amsterdam and Barcelona.

125,000
Downloads in
first month

1
Million views in
two months

#1
iTunes ranking in
several European cities

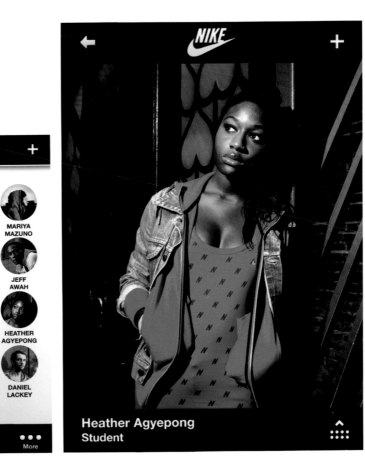

Heather Agyepong
Student

Alice for the iPad

"It's going to change the way kids learn."
<u>Oprah Winfrey</u>

When the procession came opposite to Alice, they all stopped and looked at her, and the Queen said severely, "Who is this?"

The Brief

Create an iPad version of the 150-year-old *Alice in Wonderland* story written by Lewis Carroll and illustrated by Sir John Tenniel. The app must stay true to the spirit of the original book and use the source illustrations carved in woodblock over a century ago. The app must take advantage of the tactile feature set of the iPad: all scenes must respond to touch, tilt and shake. The app should have the look of an ancient artifact but, behind the scenes, use the very latest in software physics-simulations.

The Challenge

As a well-loved book, *Alice in Wonderland* was always going to be a tricky adaptation. The challenge was to throw new technology at the story while maintaining the narrative and aesthetic integrity of the source material. There was also the technical challenge of re-mastering the original woodblock artwork, recolouring it and cutting characters out of scenes. Every moving character in Alice for the iPad took hours of painstaking work. The background scenery was redrawn in Tenniel's style to allow the characters to spring to life independently of their surroundings.

Client
Atomic Antelope

Credits
Atomic Antelope
www.atomicantelope.com

Awards
FWA Mobile

"What a curious feeling!" said Alice. "I must be shutting up like a telescope!"

And so it was indeed! She was now only ten inches high, and her face brightened up at the thought that she was now the right size for going through the little door into that lovely garden.

Alice was a little startled by seeing the Cheshire-Cat sitting on a bough of a tree a few yards off. The Cat only grinned when it saw her.

"Cheshire-Puss," began Alice, rather timidly, "would you please tell me which way I ought to go from here?"

500,000

iPad app downloads

1.4
Million YouTube views

20
Weeks in top 10 iPad book chart

The Solution

This was an extremely intensive Photoshop project, with lots of pixel-level reworking of the original Tenniel illustrations. Once the characters were carefully cut out of the scenes, the backgrounds were redrawn and physical properties were assigned to the character bodies using Chipmunk Physics. The simulated physics world uses software versions of levers, pins, pivots, elastic bands and pulleys. You can't see this all going on behind the scenes, but Alice does not use animations, it uses physics simulations that respond to the reader's actions just like a mechanical toy.

The Results

Alice for the iPad managed to preserve the intention of Lewis Carroll's original book. By subtly enhancing the reading experience the adaptation was magical rather than brash and digital. At first glance the book closely resembles the 150-year-old text, it's only when the reader tilts, touches or shakes the book that it becomes clear that this is a definitively 21st-century experience. Kids and adults are quite delighted when the White Rabbit's pocket watch swings about, the Queen of Hearts throws tarts across the screen, or the Mad Hatter's head bobbles insanely. The key to success with this kind of digital literature seems to be a considered artistic sensibility, and not just raw technological enhancement.

F1 2010 Timing App: Championship Pass

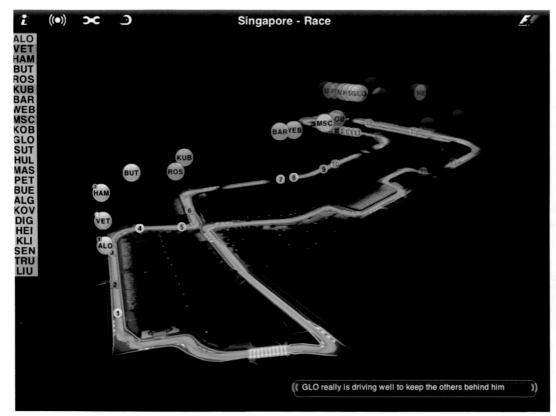

ℹ ((•)) ✕ ⌘   Singapore - Race   **F1**

ALO
VET
HAM
BUT
ROS
KUB
BAR
WEB
MSC
KOB
GLO
SUT
HUL
MAS
PET
BUE
ALG
KOV
DIG
HEI
KLI
SEN
TRU
LIU

((GLO really is driving well to keep the others behind him))

"A truly unique and engaging way to follow Formula One. It really adds another level to your Grand Prix viewing."
<u>Justin Dutton</u>, Digital Design Manager, Soft Pauer

P	CAR	NAME	LAP TIME	GAP	S1	S2
1	8	ALO	1:51.013	0.0	29.6	43.4
2	5	VET	1:50.582	0.8	29.3	43.4
3	11	KUB	1:53.619	16.7	29.9	44.6
4	4	ROS	OUT	20.0	52.1	
5	9	BAR	STOP	23.0	29.7	44.3
6	6	WEB	1:53.092	24.3	29.8	44.0
7	2	HAM	1:53.671	31.2	29.4	43.2
8	1	BUT	1:51.815	38.4	29.8	43.7
9	23	KOB	2:14.268	43.2	29.9	
10	14	SUT	1:55.574	49.8	30.1	
11	10	HUL	IN PIT	50.6	30.0	44.8
12	7	MAS	1:51.815	51.5	29.8	44.7
13	12	PET	2:14.268	52.6	30.1	44.3
14	16	BUE	1:55.574	55.0	30.3	

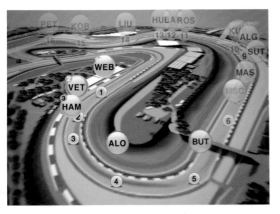

The Brief

Grand Prix in the palm of your hand. Develop a portable application that will give fans a greater insight into Formula One. Providing the user with unprecedented access to a variety of information, including the same live timing and track positioning data used by the race engineers in the pit lane.

This exclusive access will cover every event from the Grand Prix weekends, including practice, qualifying and race sessions for the entire season. Introducing this new portable service will allow fans to gain more access to Formula One keeping them in touch with what's going on while on the move.

The Challenge

The main challenge was creating an application that not only engages the massive audience of fans, but builds a high level of interest, that ultimately establishes Formula One Timing app as an integral part of Grand Prix viewing.

Delivering a vast amount of information and technical data to the portable devices was also a challenge, particularly making it appeal to a broad range of fans. The live timing data is a key element we needed to convey, in its traditional format it consists of full screens of data. We wanted to give fans access to the same timing data supplied to the teams in the original raw format, but also find innovative new ways of displaying this data and appealing to a wider audience.

03 Promotional

Client
Formula One Management
(FOM)

Credits
Soft Pauer
www.softpauer.com

Awards
FWA Mobile; SportBusiness

P	CAR	NAME	LAP TIME	GAP	S1	S2	S3	PIT	LAP
1	8	ALO	1:51.013	0.0	29.6	43.4	37.9	1	31
2	5	VET	1:50.582	0.8	29.3	43.4	37.7	1	31
3	11	KUB	1:53.619	16.7	29.9	44.6		0	31
4	4	ROS	OUT	20.0	52.1			1	31
5	9	BAR	STOP	23.0	29.7	44.3		0	31
6	6	WEB	1:53.092	24.3	29.8	44.0		1	31
7	2	HAM	1:53.671	31.2	29.4	43.2		1	31
8	1	BUT	1:51.815	38.4	29.8	43.7		1	31
9	23	KOB	2:14.268	43.2	29.9			0	31
10	14	SUT	1:55.574	49.8	30.1			1	31
11	10	HUL	IN PIT	50.6	30.0	44.8		1	31
12	7	MAS	1:51.815	51.5	29.8	44.7		1	31
13	12	PET	2:14.268	52.6	30.1	44.3		1	31
14	16	BUE	1:55.574	55.9	30.3			1	31
15	3	MSC	1:51.815	55.9	30.6			1	31
16	17	ALG	2:14.268	55.9	30.9			1	31
17	22	HEL	1:55.574	55.9	30.3			1	31

060:25
32/61 | 3:KUB 4:ROS 5:BAR 6:WEB 7:HAM 8:BUT

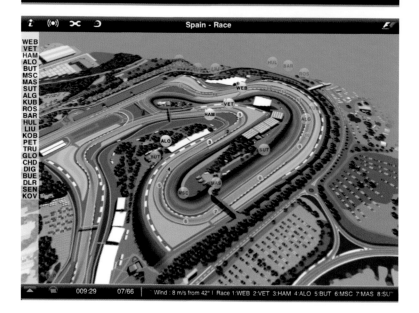

Spain - Race

WEB
VET
HAM
ALO
BUT
MSC
MAS
SUT
ALG
KUB
ROS
BAR
HUL
LIU
KOB
PET
TRU
GLO
CHD
DIG
BUE
DLR
SEN
KOV

009:29   07/66 | Wind : 8 m/s from 42° | Race 1:WEB 2:VET 3:HAM 4:ALO 5:BUT 6:MSC 7:MAS 8:SU

The Solution

Using the latest touchscreen devices we created a highly interactive and dynamic experience that really engages the user. Along with news, results and stats the application streams real-time data directly from the race track to the palm of your hand.

The features include a live timing screen giving fans a chance to analyse the drivers' and teams' performance throughout the race weekend, and an interactive 3D track map showing real-time positions of every driver circulating the track, exactly as they are in the race. You can zoom in and out, and pan around instantly putting you at the heart of the action.

The app's interactive map combined with the other features makes it much easier to understand how the race is unfolding, offering a new dimension to Formula One.

The Results

With a feast of live information and data available at your fingertips, Formula One Timing app has achieved unprecedented results since its launch as a portable service. The fan base has grown year on year and support for the app has been overwhelming throughout the Formula One community.

The quality and innovative display of data, together with many unique features, has meant it's not just the fans who have taken to the app, but professionals alike, who use the app to receive updates in the commentary box while broadcasting live to the nation. The app has become a valuable companion to Grand Prix viewing, allowing fans to get closer to Formula One than ever before.

Pos	Car		Driver	Lap	Lap Time
1			Webber	3	1:28:430
2			Vettel	3	1:28:615
3			Hamilton	3	1:28:511
4			Alonso	3	1:28:930
5			Button	3	1:29:223

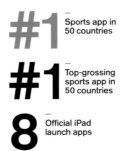

#1 Sports app in 50 countries

#1 Top-grossing sports app in 50 countries

8 Official iPad launch apps

myFry

"I think that in five years' time people
might find the archive of this interview
and find it extraordinary that we
even questioned or found amazing
that there should be such a way
of reading. I love the idea of that!"
Stephen Fry, on ITN News

The Brief

Stephen Fry's enthusiasm for technology is famous and well-documented, so with the publication of the second volume of his autobiography, *The Fry Chronicles*, Penguin was looking to stretch the book's boundaries. The brief? To create "the future of reading" on a handheld device.

Through myFry, Dare and Penguin sought to redefine the broader definition of a "book" by creating an entirely new reading experience – one that allows readers to delve in and out of a book by topic, and to consume a work in bite-size chunks in a non-linear manner.

A "book" is no longer just a stack of pages bound inside a hardback or soft cover. Instead, in the digital age, a "book" needs to change with the times, with its readers, with their habits and with their capacity to be entranced by new technology. This change has been evolving over time, with audio books and with ebooks, but our brief was to take the innovative reformulation of the book to new and exciting levels.

Client
Penguin Books

Credits
Dare
www.daredigital.com

Awards
UK App Store #1; Best App of the Year 2010, The Guardian

The Challenge

The launch of *The Fry Chronicles* already included four variants of the "book": hardback, ebook, enhanced ebook with video, and audio. The challenge was to deliver a mobile app that not only gave readers a completely new way of consuming *The Fry Chronicles*, but was also completely different from anything that had gone before. Penguin and Dare had three audiences in mind.

First, the fans. Stephen Fry is the closest you can come these days to a national treasure, and his fan base is both wide and deep.

Our second audience was Stephen Fry himself. With Stephen being an advocate of social networking and the latest gizmos, we ultimately had to design a product that the man himself would enjoy enough to include in his promotion of *The Fry Chronicles*.

Third, the book is aimed at people looking for a novel and creative way to access a book – be they Stephen Fry fans or not – and in so doing help to broaden the definition of what "reading" is. That's because not everyone wants to devour a book from first word to last, but may instead want to nibble. And because not everyone has hours to spend with books, myFry allows readers to finish a section on a 20-minute train ride.

03 Promotional

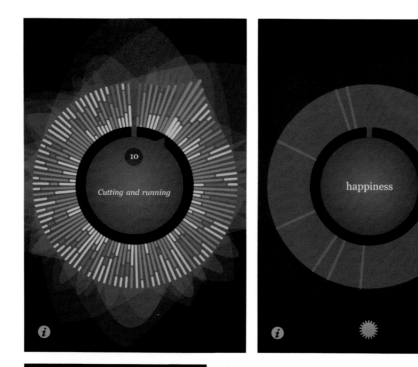

◀   **48.**   ▶

Reading: comedy

Hugh was determined that the
Footlights should look grown-up but
never pleased with itself or, God
forbid, cool. We both shared a horror
of cool. To wear sunglasses when it
wasn't sunny, to look pained and
troubled and emotionally raw, to pull
that sneery snorty 'Er?!! *What?!*' face
at things that you didn't understand
or from which you thought it stylish
to distance yourself. Any such arid,
self-regarding stylistic narcissism we
detested. Better to look a naive
simpleton than jaded, tired or world-
weary, we felt. 'We're *students*, for

The Solution

Penguin and Dare started with the device itself. The app would have to be quick and easy to use on a touchscreen interface, and be able to fill a minute or many with a bit of Fry brilliance. We took the traditional art of the index, and brought it sharply into the 21st century on an iPhone, making a visual display of *The Fry Chronicles* as intelligent, playful and fun as the man himself.

At the heart of the idea is a tagged colourwheel, devised by infographic artist Stefanie Posavec. Forget beginnings and endings. By spinning the wheel readers can explore the book's contents in unexpected ways, and create their own personal narrative.

We worked together to divide the book into 112 self-contained sections, each of which had at least one of the four colour-coded tags: people, themes, feelings and "Fryisms". So readers could now use the tags to explore the many varied threads that have made up Stephen Fry's life so far. From friends Hugh Laurie and Emma Thompson, to his addiction to sugar and cigarettes. From love and celibacy, to coinings like "bourgeoisi-fied" to "badolescence".

And just as the index has been visualised, so have other features of the printed book. What were once well-thumbed pages are now "favourites", and the familiar turned-down corner is now marked "read". Ideal for on-the-go bite-sized reads; this really is reading for the i-generation.

The Results

In its first week of sales, myFry outsold Penguin's former bestselling app by 100 percent, despite costing £5 more than its predecessor.

It went straight to the top of Apple's App Store charts in the UK, was the number one highest grossing of all paid apps, and is still in the all-time top 30 of all paid apps. This is even more impressive when its premium price is taken into account (£7.99 versus the £1 or less charged for most other apps).

It wasn't just the reading public – the press loved it too, with great reviews across the broadsheets, TV, newspapers and hundreds of blogs.

To prove that myFry and other innovations truly can redefine the "book", but not replace it, *The Fry Chronicles* quickly vaulted to number one in all five formats – the hardback book; the ebook; the enhanced ebook including eight new videos of Stephen Fry expanding on the work's anecdotes, with links to relevant websites and web content; the audio book; and the myFry app.

More generally, Penguin's investment in digital format books has proved to have been a prudent move for the company – and the myFry app has been cited in press reports and elsewhere as evidence that Penguin "gets" digital. Despite the current economic climate, Penguin has reported a surge in sales, with ebooks offsetting the more sluggish performance of traditional print formats.

#1
UK App Store

30
Top 30 all-time top-grossing apps

100
Percent outsold Penguin's former bestseller

The Rules of Golf

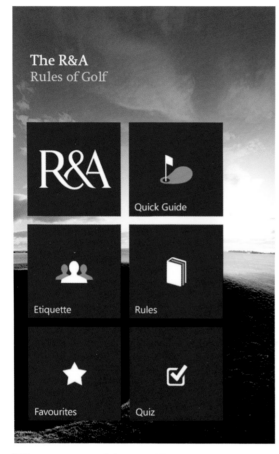

"The app provides golfers around the world with the ability to easily consume and reference one of sport's most widely read rules books. The Rules app has numerous features that explain all aspects of the laws of the game from etiquette through to the complete Rules of Golf."
Client, The R&A of St Andrews

Client
The R&A of St Andrews

Credits
Lightmaker
www.lightmaker.com

Awards
IAB

The Brief

The R&A is steeped in over 250 years of history as one of the oldest golfing bodies; they are also a modern, progressive and mobile organisation. The larger brief was to develop a strategy and digital channels to galvanise the brand across their global platform, part of this brief was to modernise the way the Rules of Golf are consumed.

The R&A tasked us with taking the Rules of Golf and mobilising them across applications for iOS (both iPhone and iPad), Android and Windows Phone 7. To take a flat and legally orientated reference book and bring it to life on a mobile digital format, leveraging the innate attributes of each platform and hardware device to augment the consumer experience and bring them closer to the brand and its patrons.

The Challenge

Developing applications across multiple platforms has its challenges but before we were to cover these, we needed to work out how we would make the content interactive – there would be little point in simply applying the text and images from the publication and digitising them into an app.

We had the source content (text, images, data and video) to bring together as well as the various cross references between sections and content types. We needed to find a way to develop the applications so that the complexities of the content were invisible to the user.

The challenges were architecture, features, usability, interaction and interface. Getting these right would give the consumer a consistent and logical experience no matter which device they were using.

03 Promotional

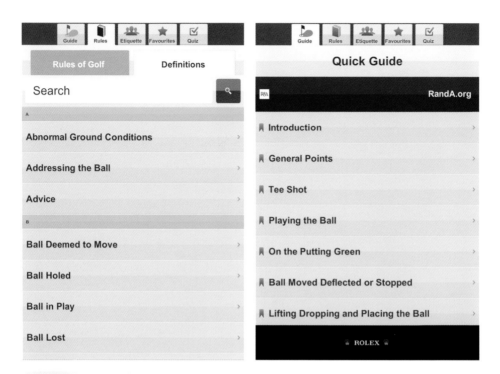

Rules of Golf | Definitions

Search

A

Abnormal Ground Conditions

Addressing the Ball

Advice

B

Ball Deemed to Move

Ball Holed

Ball in Play

Ball Lost

Quick Guide

RandA.org

Introduction

General Points

Tee Shot

Playing the Ball

On the Putting Green

Ball Moved Deflected or Stopped

Lifting Dropping and Placing the Ball

ROLEX

‹ Back **Guide**

Tee Shot

Play your tee shot from between, and not in front of, the tee-markers.

You may play your tee shot from up to two club-lengths behind the front line of the tee-markers.

If you play your tee shot from outside this area, in match play there is no penalty, but your opponent may require you to replay your stroke; in stroke play you incur a two-stroke penalty and must correct the error by playing from within the correct area.

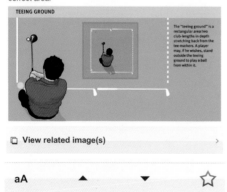

□ View related image(s)

aA ▲ ▼ ☆

100,000 Downloads
40,000 Lines of code
165 Classes
120 Articles

The Solution

We worked on a consistent IA that would fit all platforms and then wireframe concepts for the devices starting with iOS – we knew if it worked here it would adapt for the others. From here, we adapted for other devices taking into account the best native features from a navigation and interaction perspective.

Then we developed the interfaces based on the interaction model, adapting it where necessary but maintaining brand ubiquity. With the interfaces in place, we applied the feeds and content; the content was identical throughout apart from the larger assets needed for the iPad. We also included a quiz and social sharing tools within the app to appeal to a younger audience. The results were stunning and the next big task was testing, testing, testing!

The Results

The Rules of Golf for iPhone and iPad were launched during the Open Championship in July 2010 with the other platforms coming online shortly afterwards. Apple promptly featured it as iPad App of the Week and the top free Sports featured app. The application was also fortunate enough to win an IAB award in August 2010.

The application is downloaded and used every day by golfers, officials and federations and has been featured in many golf and sports publications. It's spreading the knowledge about the rules in an imaginative and enjoyable way across the globe while helping attract young players to study the rules in their more palatable and mobile form… it's quite cool!

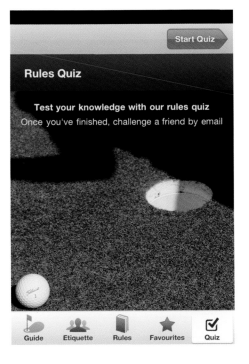

Porsche G-Force

"The G-Force app makes intelligent use of
iPhone capabilities and delivers an experience
true to the brand and the product."
Deniz Keskin, Manager of Internet, eBusiness
& Dialogue Marketing, Porsche AG

Client
Porsche AG

Credits
Fi
www.f-i.com

Awards
FWA Mobile

The Brief

Porsche approached Fi with the goal to create the world's best iPhone application leveraging the device's capabilities and linking to their world-renowned automobiles in a meaningful way. The app had to be engaging for Porsche enthusiasts and reflect the brand's cutting-edge performance and precision characteristics for their automobiles.

The Challenge

The Porsche mobile app had to stand out among existing applications released by other car manufacturers both visually and technically. It was decided early on that the app's function-ality would be linked to the performance characteristics of Porsche automobiles. The teams brainstormed to come up with the best use of the iPhone's built-in features. In order to achieve the goals required to deliver an exciting experience to the end user, we collectively decided to create an app that enabled users to track the G-forces they experienced while driving the high-performance cars.

03 Promotional

2.6g

Distance 47 ft

Speed 65 mph

185 Daily average
downloads

7.2 Gs: Highest
G-Force recorded

The Solution

Competing products from other car manufacturers calculated the G-forces in an entirely different way, providing a much less accurate experience for the end user. Fi proposed a solution that combined both the iPhone's core GPS capabilities with its built-in accelerometer to deliver results with maximum accuracy. Along with measuring G-forces, the app stores a history to provide a record of the driver's exploits. Also included is a news feed directly in the app for Porsche lovers. Fi included unique graphic details and polish that matches the sophisticated design Porsche is known for around the world.

The Results

The final product mixed beautiful design with a functional application that truly serves the Porsche customer. The client was very happy not only with the features and the design, but the level of precision the application offered, mirroring the brand values of Porsche. The app was also co-sponsored by Mobil 1, providing Porsche with additional advertising opportunities. In turn, this allowed Mobil 1 to align itself with a great brand in Porsche as well as a great product in the Porsche G-Force app.

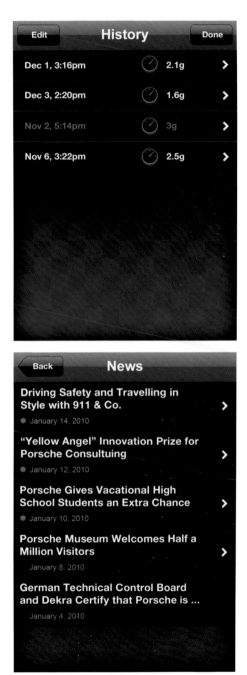

_iamChris:
Mobile Portfolio

loading...

"I love your clean approach to design."
<u>Chino</u>, Nostalgic Productions

Client
Christoph Kock

Credits
Christoph Kock
www.christophkock.com

Awards
FWA Mobile

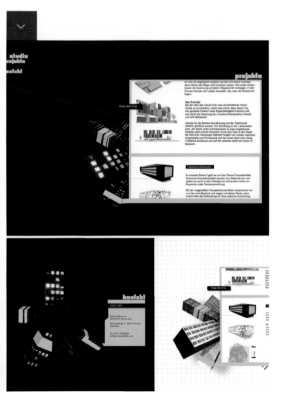

The Brief
It is difficult to showcase your work across a mobile webpage, because most people do not own a smartphone and therefore do not realise the trend and importance of it. But my will to work in the mobile segment was so big, that my own portfolio had to be used.

Although I knew that not a lot of people would visit my portfolio with their smartphones, I still decided to go ahead. Which meant, of course, working without payment and working in my free time.

On the other hand it meant freedom during work. No customer who keeps on influencing you, no appointment pressures, no arrangements etc. For me it was important that my mobile portfolio is very plain and clear and conveys only the most important content. Totally according to the motto "Less is more".

The Challenge
The most challenging part for me was the implementation of the programming, since I am originally a graphic designer and not a classical programmer. Should it be an app or better a browser based website? Which of them can I actually implement? All such questions I first put aside and started with the part with which I have the most fun, the design. But the design too had a challenging part. A small screen, no roll-over effects, big buttons for the fingers and especially fewer images to keep the size small. It must also work in landscape mode and so on.

The Solution

The solution for me was a browser based site, which gave a lot of advantages and is totally sufficient for a portfolio. Because I already had previous knowledge of HTML and CSS, I quickly decided to use HTML5, which made the entrance into the programming easier and which will be for sure the future in the mobile segment.

With a little JavaScript I also assembled small animations and crossovers. A welcoming text in a small speech balloon slides along with the content, so that a little motion is introduced into the site. But this can be removed any time with a click. The content was reduced so only the best and only part of my work is presented.

The Results

Because the site has no commercial background, my efforts to promote it were kept within a limit. But after I saw that a new category for mobiles was being given on FWA, I had to try.

The results were better than I had expected. The site won an FWA Award and it was printed in Page magazine with a small interview. Furthermore, the site was posted on many blogs and other websites. Through that the visitor numbers have risen considerably and more and more requests and feedback came in from all over the world.

7

— Days creation time

3,500

— Percent site visit increase

HoloToy

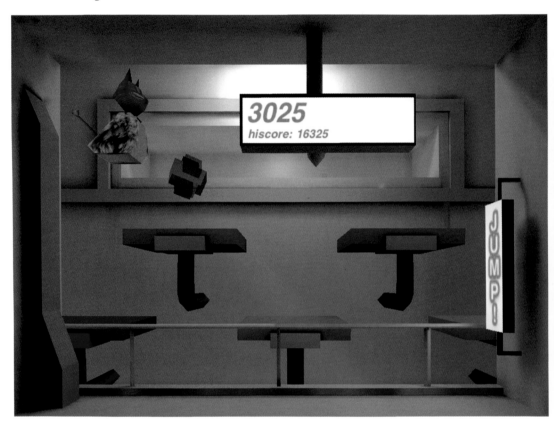

"I believe that HoloToy is really something special. I don't think you'll find a better 3D app in the whole store."
TUAW

The Brief

For as long as I can remember I have been fascinated by optical illusions and the child-like wonder they inspire in people of all ages. At the top of the list is the art of anamorphosis and trompe l'œil (French for "deceive the eye"). Both are forms of image distortion with early examples dating as far back as the murals of Pompeii. With an understanding of perspective, artists were able to create shocking 3D illusions that appeared to leap from their canvas when looked at from a certain view point. My goal for HoloToy was to create a digital interactive toy that explored this concept using modern technology.

The Challenge

Calculating the camera, off-axis projection and accelerometer filter that control the illusion took quite some tweaking before the balance felt right. One of the greatest challenges was coming up with novel uses for the tech that would make up each of the many updates along with all the new assets required which I was creating solely by myself. I wanted each update to be its own mini event and really tried to outdo myself with each new idea. Being just one guy, the frequent updates also meant that I was regularly dealing with the iTunes connect submission process, writing press releases and doing guerrilla-style promotion.

Client
Ben Hopkins, kode80 LLC

Credits
Ben Hopkins
www.kode80.com
HoloToy Icon Design
Josh Corliss
www.jayemsee.com

Awards
FWA Mobile; Gizmodo Apps of the Week; App Store Staff Picks Japan; #1 Paid Entertainment app in USA, UK, France, Japan and many more

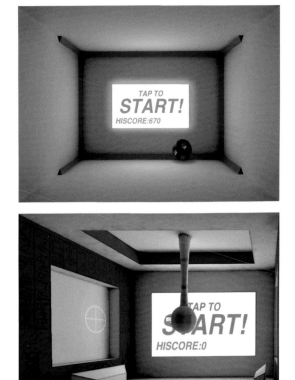

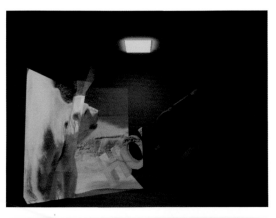

The Solution

By building a simple framework that encapsulated the core functionality of dynamic off-axis projection controlled by the accelerometer (and later gyroscope and facial recognition-based head tracking), I was able to rapidly prototype new interactive holograms which would then be deployed as app updates. Along with the code base, I also created Blender templates of the holo-room so that creating new 3D assets of the correct dimensions with consistent radiosity lighting was a relatively streamlined process. My marketing strategy was simply Twitter, YouTube and press releases centred around each new update to the app. The frequency of app updates created a steadily building cache of HoloToy-related press and media online which greatly helped with exposure and SEO.

The Results

The very first release had zero promotion and results for the first month were modest. However, by week five things really started moving. With each update, the amount of press increased, more people became aware and the features of the app itself matured. HoloToy far exceeded my expectations in the long-term and really confirmed for me that the release-quickly, update-often model of development is not only suited to the iOS eco-system but can be very successful for independent developers in general. At time of writing, HoloToy contains numerous games, interactive toys, the ability to create holograms from user photos and remains in the top charts of multiple countries around the world.

550,000 YouTube views

65,000 Sales

193,118 Upgrades

Sevnthsin Mobile

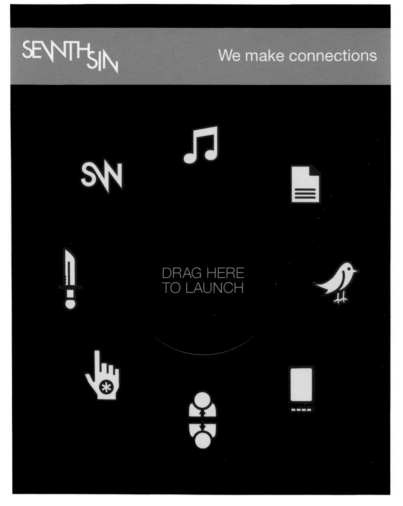

"Interaction is their game and they've brought it to the mobile web with fashion. The circular navigation is beautiful to look at and use. Once you've gone deeper than their home page, the interaction continues. Great example of interaction... definitely a must use."
cssiPhone.com

The Brief

As an interactive design agency with a focus in mobile, we wanted the Sevnthsin mobile site to not only be an intuitive guide to the Sevnthsin style and personality, but to push the boundaries of mobile design as we knew it. Basics like "don't make buttons smaller than a fingertip", and, "too much text in a mobile browser is the worst", are already tried and true best practices, but we intended to go beyond best practices.

The Challenge

Total and complete power can be a scary thing (Darth Vader, anyone?). In a way, the limitations and goals of client projects provide guidance to let you know you're on the right track. When you're your own client, limits are out the window and the desire to plaster lightning bolts and silhouettes of people playing air guitar all over everything can start to take over. This may be a uniquely Sevnthsin issue, however, it illustrates a fine point: keep your focus. Our greatest challenge was to create a real-world example of our own unique approach to mobile design without getting lost in a sea of options. It had to showcase our creativity, work and persona, and it needed to do so in three taps or less.

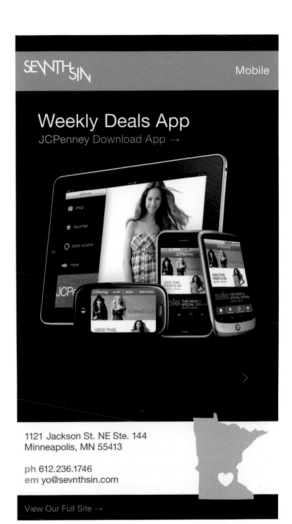

03 Promotional

Client
Sevnthsin

Credits
Sevnthsin
www.sevnthsin.com

Awards
FWA Mobile

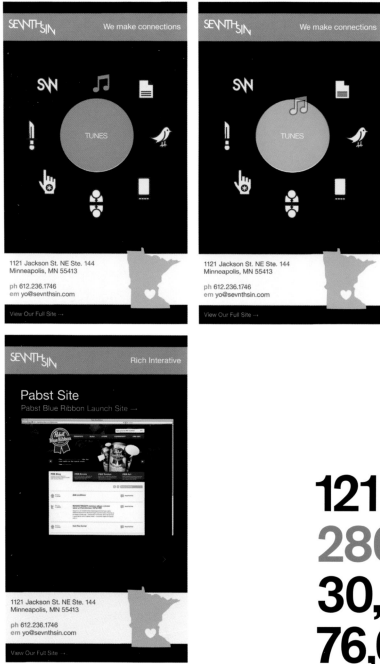

121 Countries globally

280 Referring sites

30,000 Unique visitors

76,000 Page views

SEVNTHSIN

Blog

Sevnthsin makes theFWA Mobile Site of the Day

Posted on September 22nd 2010 by JAMEY!

TheFWA is one of the most prestigious award sites around. Every day they pick only the best of the best to showcase on their site, and have been doing so for YEARS! We here at Sevnthsin are honored to have been included as part of their newest Mobile initiative by being named Mobile Site of the Day for September 22nd, 2010!

Check it out either on your Desktop Browser or your Smartphone

1121 Jackson St. NE Ste. 144
Minneapolis, MN 55413

ph 612.236.1746
em yo@sevnthsin.com

View Our Full Site →

The Solution

Our solution in five words or less: design for the device. All of our objectives and challenges were met by testing ideas against this mantra. Take our goal to let that sparkling Sevnthsin personality shine through. Instinct may compel us to write four pages detailing the origins of the company, peppered with offbeat humor and tech-speak verging on the nerdy. However, considering the device we were designing for, that much copy was out of the question. Instead we parsed feeds of our company Twitter and Last.fm accounts for personality in concise doses. Inspired by the iPhone's distinctive user interface, we forwent traditional single-tap navigation for a less conventional drag and drop approach. As for curbing our impulse to put lightning bolts on every page, lightning causes power surges that could ruin the Internet as we know it, so that answers that.

The Results

Sevnthsin mobile has been showcased on over ten design blogs as an inspirational mobile site, including SmashingMagazine.com and cssiPhone.com. Since launching the latest design iteration, traffic to our mobile site has increased 60 percent and we're averaging 151 page views a day. Overall, we'd deem our venture into the far corners of mobile design quite successful. 30 percent of the site's visitors are return visitors. Whether they keep coming back to see the latest project we've launched into the far reaches of the Internet, or simply because the navigation is too fun to play with just once, we can't always say. Either way, we're happy they stop by.

Juke Power-Up

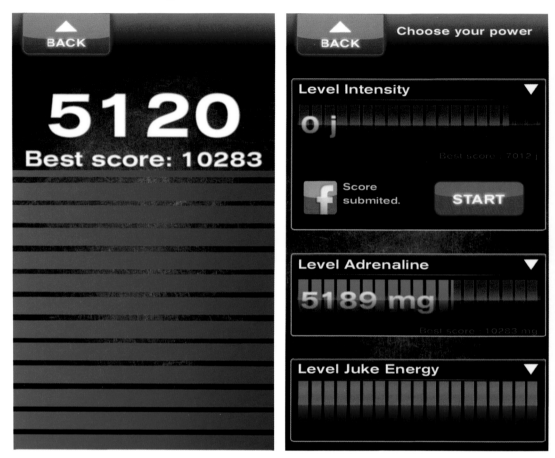

"We believe we're shifting the boundaries of how the public can download content to their phones and pioneering future industry development."
Gareth Dunsmore, Digital Manager, Nissan Europe

The Brief

The brief asked Digitas to look at how the infectious energy of Juke could be brought to life in a way that would be innovative and consumer-centric in order to support Nissan's brand and the Juke launch. At the time we were exploring browser capabilities with Flash Player 10.1 and the performance of the Android phone with it. We then received a development kit from Adobe containing a CS5 suite and a Google Nexus One.

The Challenge

The challenge was providing the same experience as an application in the browser using Flash for mobile. When the project started in early July, both Flash Player 10.1 and Froyo were in beta stage so installing Flash on the Google Nexus One was a bit tricky. Furthermore, no documentation was available for Flash or for integration. FWA Mobile didn't exist yet but it was near to the submission date of mobile content so we absolutely couldn't miss the opportunity to be the first to submit a Flash-based MOTD. The last challenge was to be the first MOTD for mobile content.

03 Promotional

Client
Nissan Europe

Credits
Nissan Europe
www.nissan-europe.com
Mind The Gap
http://mindthegap.digitas.fr
Digitas France
www.digitas.fr

Awards
FWA Mobile

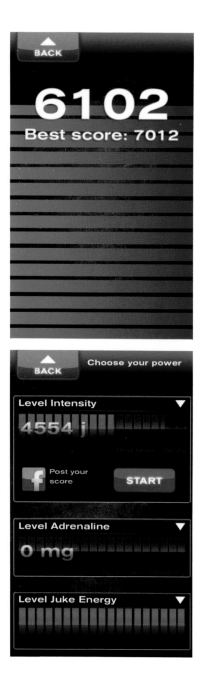

The Solution

By exploring the new API offered by Flash 10.1 we found that we could retrieve data from the Accelerometer into Flash. We then had the idea of a power ball application; this idea was entirely in keeping with the Juke concept of compact energy. However, integrating Juke Power-Up like an "application" seamlessly into the browser was not possible, neither was it to detect phone orientation. We absolutely had to lock the application in portrait mode so we used a little trick to detect phone orientation and rotate the content accordingly. This was bugged from time to time. In the end, we discovered that video was always playing in landscape mode so we used a hidden video object to lock orientation and rotate our content.

The Results

We successfully built a fun Juke experience in the browser in just two weeks with the exact same behaviour and look and feel as a real application. We liked the idea that anyone around the world with a Flash-enabled device could try out our web app.

We were the first ever Flash-based application awarded by a MOTD. The app was featured in the Adobe newsletter and in hundreds of blog articles. "Flash could kill App Store says car giant Nissan". *News of the World*. "Nissan creates first Flash app for mobiles, expect ad bombardment". CNET UK. "The first Flash-based app around". Pocket-lint.

10,000 Page views

106 Countries visiting

2 Weeks creation time

Sweet Talk

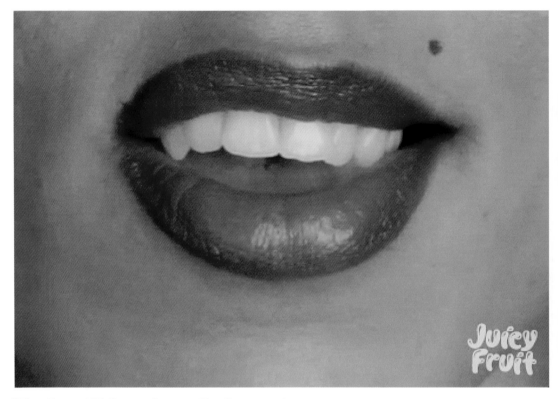

"The Sweet Talk app is an effective way to engage our target audience. Juicy Fruit fans treat it like social currency and once they discover this app, they love to talk about it and share it with their friends."
Paul Chibe, Vice President and General Manager of Gums and Mints, William Wrigley Jr. Co.

The Brief
The core teen audience for Juicy Fruit described it as "a brand on its way out" and "not relevant in my life" and the brand's promise to deliver a one-of-a-kind sweet taste seemed dated and old fashioned to these chewers – pushing its brand loyalty quite low. Teens were willing to try the sweet flavored gum; they just weren't willing to stick with the flavor exclusively. We needed to create engagement, brand buzz, and improve attitudes toward this sweet, iconic product.

The Challenge
The challenge was to rejuvenate the Juicy Fruit core brand promise of delivering sweetness. We were tasked with taking a heritage brand that was in steady decline and making it fun and exciting to a fickle teen audience. Furthermore, the challenge lay in doing so exclusively using digital media.

Client
Wrigley's Juicy Fruit

Credits
Evolution Bureau
www.evb.com

Awards
FWA Mobile

1.25 Million downloads

7 Million visits

7.02 Minutes average time on app

The Solution

We resolved to remind our teen audience that Juicy Fruit delivers the sweetness they crave like no other through a campaign that would cater to their passion points – social networking, entertainment, and mobile gaming. With the insight that teens view more videos and spend more time viewing videos than any other demographic, and that iPhones and iPhone applications have become a type of social currency for them, the free Sweet Talk application was born. An eclectic cast of five entertaining personalities was created from which users could select 25 sweet sayings, hold their iPhone up under their nose, and let the sweetness fly. Users can share a Sweet Talk message via email or Facebook straight from the app.

The Results

The viral loop and high entertainment value, along with easy sharing features built into the app, made for a highly successful application as measured by downloads and media buzz. From its introduction on July 28, 2010 through November 28, 2010, the Sweet Talk iPhone app has been downloaded 1,246,869 times and Sweet Talk videos have been played 23,215,690 times. In its second week of being available, Sweet Talk was the number one downloaded free entertainment app in the iTunes store and the number five downloaded free app overall.

iPad Touching Stories

"It's exciting to see agencies get to grips with what is possible on this device, and it shows just how much engagement can potentially be attained with your next advergame, branded content series or even a banner."
Contagious magazine

The Brief

In early 2010, Boards Magazine commissioned Tool and Domani Studios to create something cool that would be featured in the cover article of their biggest issue of the year, the "It" issue, which was to be released at the most high-profile advertising event of the year: the Cannes Lions Advertising Festival. Boards wanted something experimental… something that explored the future of interactive storytelling on the iPad. They wanted us to encourage and inspire other content developers and technologists.

The Challenge

The goal was to catch the early adaptors while they were still figuring out this new technology and give them four very unique and different perspectives into how iPad-specific storytelling might evolve.

What could you do with iPad storytelling? How might you leverage the native features of the iPad such as multi-touch, the accelerometer and Internet browsing within the video experience?

We really wanted to excite people about the possibilities, but at the time, the iPad was so new, we were like everyone else; we were trying to quickly learn all the features so we decided to create prototypes before we started constructing the stories.

Client
Tool of North America
Domani Studios

Credits
Tool of North America
www.toolofna.com
Domani Studios
www.domanistudios.com

Awards
FWA Mobile

9,000 Downloads
1,000 Tweets

The Solution

Five Tool directors — Sean Ehringer, Erich Joiner, Tom Routson, Geordie Stephens, and Jason Zada — along with technology partner, Domani Studios, concepted, shot and developed four interactive, live-action, short stories that are designed specifically to take advantage of the unique, interactive features of the iPad.

Each story was intended to show off different interactive storytelling mechanisms (e.g. shaking of iPad, Internet browsing). The four stories were packaged together in a single, free app and released in the iTunes store. We decided to title the app what we'd been calling the project, internally — Touching Stories. And by touching, shaking and turning the iPad, the user can navigate, unlock and reveal unexpected variations in each of these stories.

The Results

Upon its launch, Touching Stories was featured in a one-week exclusive in Fast Company. Each day a different story was featured with creative insights from each of our directors. It's been covered in international publications such as *Creativity, Contagious, psfk, Shoot!* and *Computer Arts*. Most recently, it's been selected as one of the first "Mobile of the Day" awards by the FWA.

choose a response

"I don't have time for riddles, old man."   "Help me, you son of a bitch."   "Where is she? I need help. Please."

pinch with your fingers to grab the keys

Nike Football+
Team Edition

"As both a coach and a player, I love this app. I am 34 now and it's been years since I've had any formal coaching. This app gives me and my teammates drills that we can do on our own time to improve every aspect of our game."
Adam Johnson, App Store user review

The Brief

Nike and AKQA were given the exciting opportunity to create an app that coincided with the launch of the iPad. Nike briefed the agency to create an app that would – in the strong tradition of our training applications – bring to life Nike Football's mission of delivering inspiration and innovation to every footballer in the world. We were asked to maximise the iPad's video-rich characteristics and bring something new to the digital football world.

The Challenge

To help coaches get the most out of their team, we developed a fully customisable digital football training tool to use during training sessions. Nike Football+ Team Edition is designed for coaches of all levels to use throughout their season and pre-season. It gives a coach the tools to improve a squad's Speed, Control, Accuracy and Fitness – individually and en masse – using instructional drill videos, challenges and pro insights on the iPad.

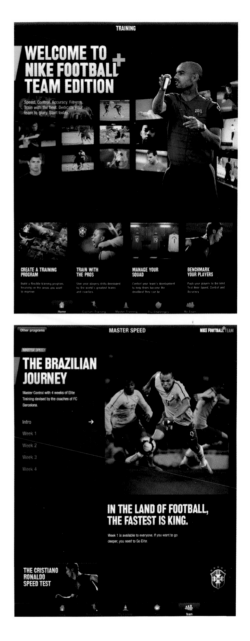

Client
Nike

Credits
AKQA
www.akqa.com

Awards
FWA Mobile

The Solution

Nike Football+ Team Edition enables coaches to build and customise bespoke training programs for each of their players, focusing on specific areas of improvement. The application brings pro drills created by the world's greatest teams and coaches straight to the iPad. Teams can Master Speed with Brazil, Accuracy with Juventus, and Control with Barcelona. The team can be put to the ultimate test, taking on the world's best players to measure their skills in the Master Speed: the Cristiano Ronaldo Challenge, the Fernando Torres Accuracy Challenge and Cesc Fabregas Control. Every player has a record of his test scores. The coach can then simply assign the relevant video drills to each player depending on his strengths and weaknesses.

The Results

The app helped to achieve brand objectives by making young footballers better by providing their coaches with a state-of-the-art training tool. The app has been highly praised for its beautiful and user-friendly design and the way its video-rich content leverages the iPad's display. Team Edition has proved to be a very cost-effective way of targeting coaches via effective repurposing of premium training videos from online training programs.

We've created the first digital training aid of its kind – a fully customisable football coaching app exclusively available on the iPad. With the innovative capabilities of the iPad, we were able to use instructional drill videos, pro insights and challenges – all customisable and accessible anywhere to completely innovate the way coaches train their teams.

35,000 Downloads

3 Project team

6 Weeks project time

1 Nike's first iPad app

Everything Falls

"They have produced an extremely satisfactory application through a combination of an excellent capability in planning, materialization, and commercialization of their partner's (client's) concept and the flexible and positive attitude in the communication with their UI planning and design."
Munsu Park, Deputy Manager, Woongjin Thinkbig Co, Ltd.

Client
Woongjin Thinkbig Co. Ltd

Credits
Kobalt60 Co. Ltd
www.kobalt60.com

Awards
#1 Korea iPad App Store

The Brief
It was our first project targeting the iPad, and we had to make an interactive book for children to experience the scientific phenomenon of gravity. It was a good opportunity to work with a new department set up to advance into a new media field by Woongjin Thinkbig, a quality education publication company.

The Challenge
We spent a lot of time creating an easy and intuitive user interface on a bigger screen size compared to existing mobile devices so that children could use it easily. We also had to work out dynamic elements that could not be conveyed by conventional print media. Additionally, it was an important task to select books with themes that could be produced most efficiently on the iPad.

It was a big challenge for most of our staff to make themselves familiar with new tools and technology, as they had previously been developing programs based on Flash ActionScript.

03 Promotional

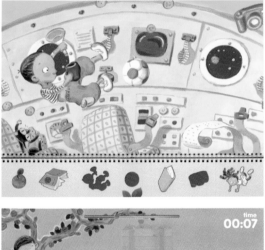

The Solution

When trying to kick off the project before iPad was made available in Korea, the company's own past project experience producing applications for iPhone was a lot of help. In addition, appropriate effects and functions were expressed with the excellent frameworks of cocos2d and box2d. Unknown errors and many problems might have gone unsolved during the project without the cooperation of colleagues and the assistance of technical books plus a lot of online searches.

The Results

The project was completed successfully by maximising the utilisation of the iPad's accelerometer and touchscreen so as to make the subject of gravity understood. Its English version was produced at the same time and has brought about a good response in Korea in terms of English education for Korean children. This has also enhanced its chance of advancing into the world market. It is with all the help we received that we have enjoyed the honour of being the most popular in the education app field in the Korean App Store.

#1 Korea App Store, Education

#1 Top paid app in Korea

20 Top 20 US App Store, Education

fborn Mobile

firstborn·

PROJECTS   ABOUT   JOBS   CONTACT

Firstborn
Under The Hood
More info

The Hartford
Achieve What's Ahead
More info

Ford Edge
MyFord Touch
More info

SoBe
SoBe.com
More info

R.M.S. Titanic, Inc.
Expedition Titanic
More info

Wrigley
5® REACT™
More info

Firstborn
Dofl Ball
More info

Wrigley
Starburst iPhone App

"At a time when interactive agencies are all about to start going mobile, Firstborn have set the bar with a mobile site for all devices."
Rob Ford, FWA

Client
Firstborn

Credits
Firstborn
www.fborn.com

Awards
FWA Mobile

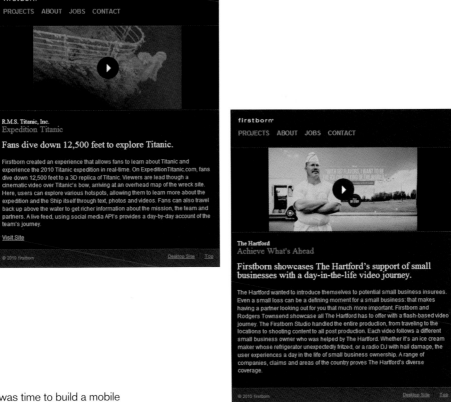

firstborn*

PROJECTS   ABOUT   JOBS   CONTACT

R.M.S. Titanic, Inc.
Expedition Titanic

Fans dive down 12,500 feet to explore Titanic.

Firstborn created an experience that allows fans to learn about Titanic and experience the 2010 Titanic expedition in real-time. On ExpeditionTitanic.com, fans dive down 12,500 feet to a 3D replica of Titanic. Viewers are lead though a cinematic video over Titanic's bow, arriving at an overhead map of the wreck site. Here, users can explore various hotspots, allowing them to learn more about the expedition and the Ship itself through text, photos and videos. Fans can also travel back up above the water to get richer information about the mission, the team and partners. A live feed, using social media API's provides a day-by-day account of the team's journey.

Visit Site

© 2010 firstborn   Desktop Site   Top

firstborn*

PROJECTS   ABOUT   JOBS   CONTACT

The Hartford
Achieve What's Ahead

Firstborn showcases The Hartford's support of small businesses with a day-in-the-life video journey.

The Hartford wanted to introduce themselves to potential small business insurees. Even a small loss can be a defining moment for a small business: that makes having a partner looking out for you that much more important. Firstborn and Rodgers Townsend showcase all The Hartford has to offer with a flash-based video journey. The Firstborn Studio handled the entire production, from traveling to the locations to shooting content to all post production. Each video follows a different small business owner who was helped by The Hartford. Whether it's an ice cream maker whose refrigerator unexpectedly fritzed, or a radio DJ with hail damage, the user experiences a day in the life of small business ownership. A range of companies, claims and areas of the country proves The Hartford's diverse coverage.

© 2010 firstborn   Desktop Site   Top

03 Promotional

The Brief
Firstborn knew it was time to build a mobile version of their Flash-based website so people could experience fborn.com in the palm of their hand. The new mobile site had to show the most important aspects of the company, including the most recent work and current open positions.

The Challenge
The team wanted to capture the distinct functionality of the website, while leveraging the platform of a mobile device. When making a mobile version of a website, you can't just recode a site, make a mobile redirect and call it a day. Firstborn had to start the mobile site development by streamlining the content of their desktop version – getting to the core of what people want when they're on the go.

firstborn

PROJECTS   ABOUT   JOBS   CONTACT

So you want to work at Firstborn? Good, we're hiring.

New York City has been Firstborn's home since 1997 when we stepped foot into the historic Film Center Building on 9th Avenue and 44th street.

Art Director
More info

Designer
More info

Internship/Developers
More info

Internship/3D Motion Designers
More info

Internship/Designer
More info

Junior HTML Developer
More info

Senior Strategist
More info

Executive Producer
More info

Producer/Project Manager
More info

3D/Motion Artist
More info

Senior Flash Developer
More info

© 2010 firstborn   Desktop Site | Top

---

firstborn

PROJECTS   ABOUT   JOBS   CONTACT

Art Director

Email your resume or work samples to jobs.designer@firstbornmultimedia.com

Coming up with ideas isn't second nature to you, it's your natural instinct. As an art director, you're passionate about great ideas, and driven to find creative solutions with concepts. You've got your process on how to develop great ideas, make beautiful designs, and you need little guidance when bringing ideas to life. You must be able to wear multiple hats, juggle projects, and excel at working in an extremely fast-paced environment. As Firstborn's newest art director, you will be responsible for working on your own creative design solutions for multiple assignments.

Required Skills and Aptitudes

- 3+ years designing interactive projects
- Exceptional design portfolio of recognizable work
- Creativity is an absolute must! You must successfully come up with effective design ideas that are visually arresting and conceptually engaging.
- An eye for detail, typography and color
- Required technical skills: Photoshop, Illustrator, knowledge of Flash and your tried and true process of creating design.

© 2010 firstborn   Desktop Site   Top

---

firstborn

PROJECTS   ABOUT   JOBS   CONTACT

Hell's Kitchen, NYC

Firstborn
630 Ninth Avenue #910
New York, NY 10036 (map)

Phone
212.581.1100

E-mail
info@firstbornmultimedia.com

Social
f facebook   t twitter
in linked in   V vimeo
flickr

© 2010 firstborn   Desktop Site | Top

3,700

Visits first month

7
Percent monthly
traffic of main site

2
Days development
time (first version)

The Solution
The first day of development was spent
building a tool to assist in updates and page
creation, speeding up the process. From there
they built a mobile version of www.fborn.com
using XHTML Mobile for broader cross-
browser compatibility and CSS3 animations
for an immersive experience on the devices
that support it. The goal was to replicate the
native functionality of the devices in the
website, so users could swipe page-to-page
as they're used to in applications and their
home screen. The result is an experience that
feels more like a mobile app than a web-page
in the browser.

The Results
M.fborn.com is an experience that feels like a
simple, engaging mobile app while showcasing the best assets from the website. In the
first month and a half alone, Firstborn mobile
had 3,700 site visits. Traffic trends indicate
there will be about 1,400 monthly visits, which
makes up seven percent of our Flash portfolio
monthly traffic.

03 Promotional

How Rocket Learned to Read

"We are so excited to see the book evolve into something new and exciting, without losing the focus on the story. Each page was injected with life – a unique dimension – which gives the reader a completely new book experience."
Lee Wade, Publisher, Schwartz and Wade

Client
Random House

Credits
Domani Studios
www.domanistudios.com

Awards
FWA Mobile

The next morning
Rocket arrived early.

The next morning
Rocket arrived early.

The Brief

How Rocket Learned to Read is a *New York Times* children's book best-seller. Our goal was to make the iPad application an Apple best seller. Most iPad children's book applications come standard with moving illustrations, narration, and touch-point interactions. To take Rocket to the next level, we pushed the animation and design as far as possible to maximize the iPad's most unique features and provide a truly immersive experience for children. Additionally, since learning to read is the theme of the story education was a top priority throughout development.

The Challenge

Finding an animation style that retained the charm of the book's hand-drawn illustrations in an iOS application was no small feat and considering the lack of existing animation tools within the iOS framework we knew we would need to build our own to get it right. The warmth and life the images of Rocket and Bird easily convey had to be matched in their vocalizations to keep kids interested in the characters and encourage their imaginations. Because it's a long book, we needed to balance the linear narrative with the interactivity. We wanted the structure to encourage kids to keep reading, but also pause at points to engage more deeply in interaction.

When Rocket spelled
M-U-D,
he knew that spring,
as it always does,
had returned.

and
C-O-L-D.

Rocket loved
to play. He loved
to chase leaves and
chew sticks. He
loved to listen
to the birds sing.

The Solution

Starting with an experimental mindset, we played around with prototypes and developed a new animation framework for animating on the iPad.

We discovered that animating entirely within Flash allowed our animators to use tools they were familiar with and enabled us to fine-tune the style to the client's satisfaction before getting too engrossed in code. From there we built an animation engine that was able to import these animations from Flash and dissect the scenes into objects and animation paths. Our system recreated each scene using native iOS programming rendering it identical to the animations the client had previewed in a Flash-enabled browser.

The Results

The final product has stunning animation sequences that move gracefully through the narrative and are laced with a perfect balance of educational, entertaining, and device-centric touch points. These include a sight word game, a mud-splashing page, microphone animation activation, an alphabet drop game, drawing in the snow, and Rocket and Bird sharing their story through movement.

44,000
Lines of xml

36,000
Animation key-frames

1,013
PNGs

And when they were done, they read it again.

And again.

And *A-G-A-I-N.*

Social

Introduction by
Iain Dodsworth, TweetDeck

04

The mobile phone is, and always has been, an inherently social device from the simple act of calling or texting someone to the use of a built-in address book. But our current understanding of "social" is shifting from a definition that encompassed just communication to one which now includes sharing and discovery.

"We find ourselves today in a landscape populated with more powerful devices than we've ever seen before, the likes of which feel more like science fiction than fact."

We find ourselves today in a landscape populated with more powerful devices than we've ever seen before, the likes of which feel more like science fiction than fact. As ever the pace of innovation is breathtaking, and not confined to hardware. Services and products built for these devices now assume a "social layer" as a matter of course (or are derided if it's missing) and as onlookers one cannot help but be somewhat in awe of the continuing growth of Facebook and its goal to "socialise the web". The social tentacles in all their forms are mainstream.

Putting aside the arguments and concerns of having a single powerful company at the heart of all things social, Facebook does provide a key ingredient to the continuing growth and acceptance of social media – a centralised place where we are happy to share, discover and interact.

Knowing where your audience is, in an online sense, and being able to contact them and their social groups directly is an incredible advantage for those entrepreneurs and developers building the next generation of social startups. The Facebook scale provides a foundation to the social layer and as more layers (services, products, companies) are built on top it becomes stronger.

Indeed it does seem today that socialising the web, i.e. adding a socially focused layer of functionality to websites, is an exciting and inevitable element of our online future. Whilst in total agreement with this, the truly powerful developments will start to emerge when the social layer has become so prevalent as to be effectively invisible. One can perceive a time where the global social graph, and the almost infinite connections therein, is as much a protocol as the web or email.

In introducing the benefits of social networks and driving its adoption out towards the mainstream, the major social services are also, inadvertently, incubating the new breed of social startup. A one-size service does not fit all, even one the size of Facebook. Or to put it another way, a service cannot be all things to all people – we as users always want something else and these gaps are opportunities.

Right now could be the most interesting period in the growth of social media and networks for this very reason. The hype and excitement around new social startups is greater than it's ever been and is only matched by the entrepreneurial desire to innovate in this space, to compete and to have the "first class" issues of scaling to millions of users.

"This social experimentation will continue until we understand how people get real value from 'social', what keeps them coming back for more and what is and what is not palatable to pay for."

Consolidation is inevitable but trying to predict who the winners will be is fruitless – and of course that's where the attraction lies. Do location services have enough reward to become meaningful social layers in their own right or are they simply features waiting to be copied or swallowed up? Will local deal startups continue to grow at the current phenomenal rates or will customers become desensitised due to over-exposure? Are our current forms of advertising applicable to monetising social products or is there a social monetisation holy grail out there waiting to be discovered?

This social experimentation will continue until we understand how people get real value from "social", what keeps them coming back for more and what is and what is not palatable to pay for. We'll figure out what "social" really is, how it means different things to different people and how it can differ depending on the device being used or the circumstance of the situation.

04 Social

So, what of the future? In a moment of pure guesswork, I'd like to see "social" become so embedded in the fabric of our online activity that we're no longer cognisant of its presence. The beginnings of this can be seen now in a subtle but important shift in behaviour whereby some people now check their Twitter stream or Facebook messages first before their email.

"Our devices would harness their awareness to filter, analyse and block the growing torrent of information we are undoubtedly going to be bombarded with."

I think after a period of "social layering" and "socialising the web" our reliance on social functionality would demand a different, more fundamental approach that is far removed from the idea of sticking a "social band-aid" on an ageing website. The social operating system would be born and would make our devices not only "socially" aware of each other but also spatially aware of their surroundings and context (location and time). Our devices would harness their awareness to filter, analyse and block the growing torrent of information we are undoubtedly going to be bombarded with. All the while sitting silently in your pocket, even ready to take a phone call.

Iain Dodsworth
TweetDeck

Bio.
Iain Dodsworth
TweetDeck

Iain was one of the first people to graduate from the Systems Modelling degree at Sheffield Hallam University and from there went on to spend the next 12 years working on a few startups of his own and in data visualisation for numerous financial institutions in the City of London.

In 2008 he released TweetDeck, an application designed to make consuming large amounts of social information (from the likes of Twitter, Facebook, LinkedIn, MySpace etc.) both manageable and efficient. Over the past two years he has taken the company through two rounds of funding and grown it to a team of 15 with offices in London and New York. Several million people use TweetDeck every month and it is widely regarded as the most powerful way to engage in social media.
–

www.tweetdeck.com

"Some people now check their Twitter stream or Facebook messages first before their email."

Android TweetDeck

Client
TweetDeck

Credits
TweetDeck
www.tweetdeck.com

Awards
FWA Mobile; TechCrunch
Europa award for 2010

"TweetDeck started making sense of Twitter for 'pro users' only in 2008 but has since become a mainstream tool. Now aggregating a variety of social platforms, it's extended to other devices and continues to innovate at a fast pace under CEO and founder Iain Dodsworth."
TechCrunch Europas Award panel

The Brief

Develop a social networking experience for Android mobile users that mimics the rich experience they receive on the award-winning TweetDeck Desktop client. TweetDeck has rapidly become the world's leading desktop social browser, and a compulsory tool for consuming real-time streams. Connecting you with your contacts across social networks such as Twitter, Facebook, Google Buzz, Foursquare and more, TweetDeck presents that information in a unique multi-column format. On the Android platform users should be able to see their usual columns, as well as the expected information within those columns, and at the same time have access to all of the TweetDeck features they are used to.

The Challenge

Developing TweetDeck for a mobile device has its issues, not least of which being the limitations of a smaller device and much smaller screen-size. A whole new user interface would need to be developed in order to fit the standard TweetDeck features into a handheld device, while not losing any of the TweetDeck experience. An easy way of navigating across columns would need to be implemented.

Furthermore, Android devices largely use touchscreen displays as their primary form of navigation, which is in stark contrast to the keyboard and mouse used for the desktop application. There are also technical constraints regarding memory usage which come into play, and present the developer with a few head-scratching issues that they need to overcome.

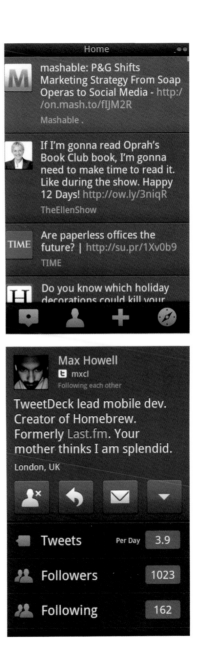

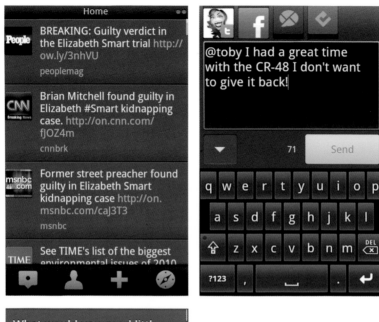

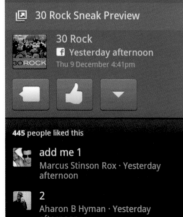

20 — Million desktop TweetDeck downloads

2 — Million iPhone downloads

4 — Million tweets per day from TweetDeck

40,000 — Searches added per day

The Solution

Through a period of design iterations, a user interface was created that perfectly combined the rich feature set of the traditional desktop version of TweetDeck with the exciting features available on a mobile device. Building the app for touchscreen input brought with it the benefits of gestures, which seem incredibly natural when moving between columns – you just need to swipe from side to side to see the adjacent columns, and then swipe up and down to scroll through the information in the column. A thorough beta-testing process with 36,000 beta-testers was put in place, generating direct feedback from users and allowing time to tweak and refine the Android app before the large-scale launch.

The Results

A high level of innovation, and an overwhelming amount of praise from Android users, and other developers in the Android community. Working on the platform from the ground up encouraged a considerable amount of creative thinking resulting in some brand new features. One example was the idea of combined columns such as a "Home" column, which aggregates all of your social network feeds, and a "Me" column, which combines all the messages that are just about you. Within six weeks of launch, the Android app made up almost one fifth of the people using TweetDeck. And after such critical acclaim from bloggers and consumers alike, the new features and UI will be rolled out into future versions of TweetDeck on the iPhone and iPad too.

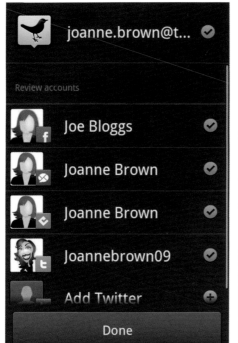

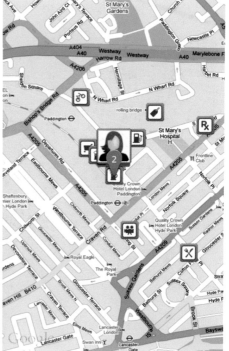

Flipboard

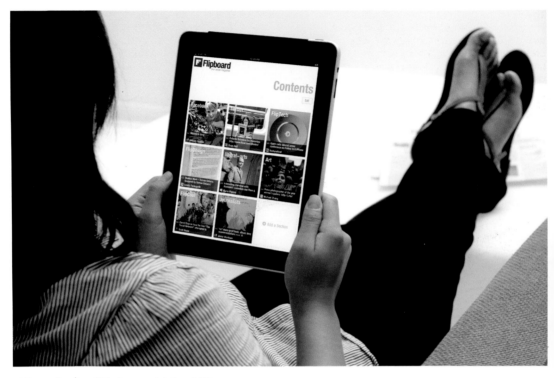

**"When I use Flipboard I feel like
I live in the future. (YES!)"**
<u>gregmcqueen</u>, User

Client
Flipboard

Credits
Flipboard in-house
communications team,
led by Marci McCue

Awards
FWA Mobile of the Month;
Apple iPad App of the Year;
Wired App Stars;
Macworld's App Gems

The Brief

The small in-house communications team was tasked with the introduction of Flipboard, a completely new company and a brand new product. The objective at launch was to establish a new way to enjoy and experience all of the content being shared between people and have Flipboard become the most popular way to discover, view and share the things being shared in social networks. Additionally, the longer-term goal was to become the "hottest" iPad app and a reason for people to buy an iPad.

The Challenge

Introducing a new product into a crowded space with many solutions that claim to "improve your social media experience" required rising above the noise level and differentiating Flipboard from any other existing product. The team didn't just have to establish a new company and a new product, but also a completely new product category. After all, Flipboard was the first company to apply the century-old principles of print to social media to offer a better way to deal with the explosion of information created by them. Additionally, it was important to explain the relevancy of this new category to people's daily lives to make them understand the usefulness of Flipboard, which was especially challenging given Flipboard is only available on the iPad, a new platform that was just emerging at the time of launch.

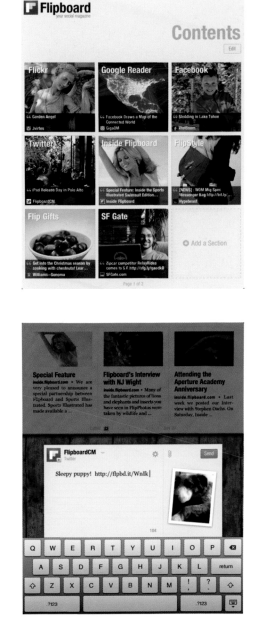

Flipboard

< FLIP

toy car

Shared by Shikhiu Ing

Recent contributors:

The Solution

The most impactful strategic decision was to position Flipboard as the world's first "social magazine" that gives people a single place to quickly "flip" through the news, photos and updates. This concise description set Flipboard apart from any other product and painted a bigger vision for the company.

The focused communications plan allowed for deeper connections with a few reporters and influencers who could help shape the story as well as the product itself. The team gave them insight into Flipboard months in advance, informed them of the $10.5 million received in funding as well as the acquisition of The Ellerdale Project, a web intelligence company. This led to thoughtful stories in media publications and on blogs, sparking online conversations on Twitter and elsewhere. This, in turn, triggered additional interest and created a mini media frenzy.

The team also worked with film-maker Adam Lisagor to bring the message to life for a broader audience, by visualizing the essence of Flipboard in a short video, making it easy for anyone to understand what a social magazine is.

The Results

Three months after the launch, Flipboard already feels like a household name (at least for iPad owners). In the weeks following the launch, the company received hundreds of requests from publishers, content owners, advertisers and other parties to work together. User adoption shows a steep growth curve and Flipboard has won an impressive list of awards and acknowledgements in just the first five months of its existence, including iPad App of the Year by Apple and being listed amongst "The 50 Best Inventions of 2010" by Time magazine.

85,000
News stories in 5 months

466,397
YouTube views

12.5
Percent estimated iPads with Flipboard

10.5
Million USD funding

The BA Foodist's Top 10 Dishes of 2010: http://bit.ly/ie5VbJ

Bon Appetit Magazine Monday   Updating... ▼

Sports Authority
& Foursquare

"Foursquare has been a great partner and their
platform has played a key role in some of our
more innovative digital and mobile campaigns."
Sean Collins, SVP Marketing, Strategy &
E-commerce, Sports Authority

The Brief

Loyalty is no longer tied to a plastic swipe card or membership club. As consumers continue to self-identify as loyal patrons of retailers, the definition of loyalty continues to change. Concurrently, the movement of commerce and retail engagement happens increasingly within the mobile environment, real-time.

In response to these two key marketplace trends, Sports Authority began their partnership with Foursquare by providing incentives for patronage at the top sporting goods retailer's 450+ stores.

The Challenge

The challenge was to go where the consumers were – in this case, mobile platforms like Foursquare – and connect with them in relevant, meaningful ways to generate awareness and increase retailer-specific loyalty.

Client
Sports Authority

Credits
Foursquare
www.foursquare.com

04 Social

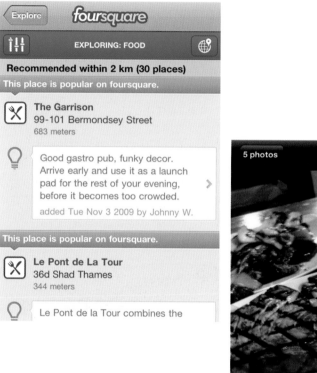

500
Percent consumer check-ins growth

20
Times black Friday check-ins

5
Million Foursquare users worldwide

The Solution
During the first phase of the Sports Authority and Foursquare partnership, the "Mayors" of each Sports Authority store were rewarded for their store-specific loyalty with a free $10 Cash Card – just for visiting the store.

In the second phase, Sports Authority shifted to a frequency-based rewards approach. Extending beyond Mayors only, all Foursquare users had the chance to be rewarded based on varying check-in frequencies.

On Black Friday (11/26/2010), Sports Authority deployed a "Check-In to Cash In" program that took online/offline integration to a new level. Consumers who checked in on Foursquare to Sports Authority stores nationwide had a chance to win one of 20 $500 Gift Cards, on the spot.

The Results
Sports Authority's consumer check-ins grew over 500 percent in a six-month period, driving measurable sales.

On Black Friday, Sports Authority saw 5x to 20x increases in Foursquare check-ins and check-ins posted to Twitter. 20 lucky Foursquare users received $500 Gift Cards and engagement across nearly every digital channel increased significantly.

KISS KUBE Mobile

"**Impressive results in a highly competitive market and in the face of a savage economic downturn.**"
The Judges, on winning NMA Entertainment category up against Nintendo and the X Factor

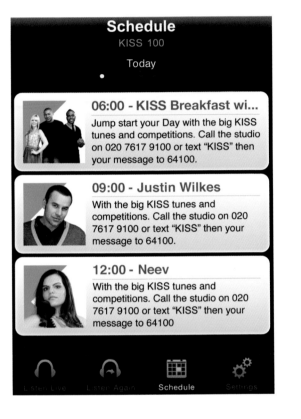

The Brief

As a famous youth music brand, Kiss has a fast-moving and tech-savvy audience, and we pride ourselves on our strength and presence on digital platforms. We wanted to provide a standout streaming experience that our listeners expect from the number one radio station for young London.

No other radio app in the UK – and in fact the world – was providing a listen live/listen again experience on mobile, not even the mighty BBC.

Streaming radio to mobile would provide us with a great opportunity to connect listeners to the brand and drive listening for the station, which is central to the radio strategy. Our audience loves mobile and they want to take Kiss with them wherever they go, but it's not enough to provide just an app that delivers a radio stream. So our objective was to deliver an app that went beyond just live radio and delivered many layers of content and leveraged social interaction.

The Challenge

Firstly and most importantly we needed to ensure the streaming experience was robust, and that the user experience was fluid and made use of all the opportunities that the touch interface offers. The listen-again experience and ability to resume from where a user left off and shuffle through audio was particularly challenging and involved a lot of clever work getting the client to talk to the server in a particular way and deliver a fluid experience.

It's not easy to deliver a deep content experience on such a small device, but we were able to fuse a great native user interface experience with a number of open-source tools that allowed us to deliver a mobilised version of our main site content in addition to the streaming experience.

04 Social

Client
Bauer Media

Credits
Synchromation
www.synchromation.com
Clicked Creative
www.clickedcreative.com

Awards
NMA Entertainment Award 2010

The Solution

We overcame a range of engineering hurdles for such a pioneering project, particularly the ability to deliver a constant stream of quality content that was lightweight enough for a mobile device, but still high-quality audio. There are a great number of behind-the-scenes functions to keep the user connected, which took a lot of work and many hours of testing! However, the small screen real estate actually provided a great opportunity to hone the message and increase interaction. Our ad formats provide click-through rates that are typically three to four times higher than standard in app display media, so this provides a great opportunity for advertisers and enhances the value of the brand from a marketing perspective; it has generated new revenue opportunities for sponsorship and promotions as well as display and pre-roll opportunities.

The Results

Kiss Kube has been a massive success, receiving 400,000 downloads since its launch in December 2009 and every month Kiss Kube mobile is used to deliver over one million streams. The app's strength lies in its under-the-hood functionality that keeps the stream flowing and switching seamlessly and uninterrupted from Wi-Fi to 3G. Kiss Kube gives a desktop-quality performance on your mobile.

Kiss Kube mobile also offers visualisation to the player, keeping listeners up to date with the latest insight, offers and competitions from Kiss. In addition, the app integrates social media tools, allowing listeners to tweet or Facebook their status whilst listening to Kiss. Users can listen to on-demand content by show, genre, name or from the last seven days of programming as well as highlighted feature shows. With device features such as swiping, users can easily navigate and access content.

1 Million streams per month

400,000 Plus downloads

KISS KUBE

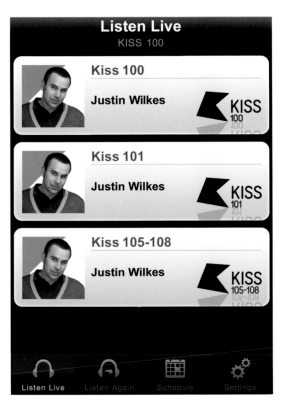

LEGO Photo

"It's elegant, well designed, and the results –
which convert your images into pretty
1x1 Lego mosaics – are pretty."
Gizmodo

The Brief

One of the misconceptions about LEGO is that it's only for creative types. While not everyone is necessarily creative, everyone has ideas. Whether those ideas are big or small, super creative or just remembering where you put your keys, they're ideas. We wanted the final product to be less about creativity and more just about having ideas and making those ideas come to life.

The Challenge

To introduce the world to the LEGO ethos of enabling anyone to create, we were tasked with crafting LEGO's first iPhone app, which would allow people to transform photos into LEGO mosaics. We knew our app needed to appeal to LEGO fanatics and casual users alike. A low barrier to entry was key. To streamline the creation process, a photo stands in for an idea – a self-portrait, a photo of a beloved pet, or a moment captured in time – and allows anyone to take that photo and turn it into LEGO blocks.

Client
LEGO Systems Inc

Credits
Pereira & O'Dell
www.pereiraodell.com
StruckAxiom
www.struckaxiom.com

Awards
2010 One Show Interactive: Bronze; OMMA Awards; AICP Next Awards; #4 US, #2 UK, #1 Japan – App Store

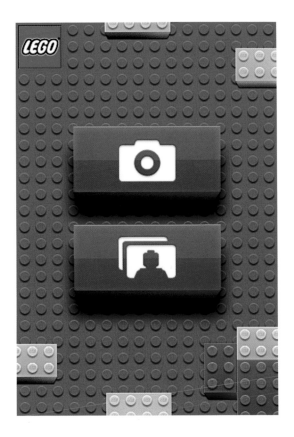

04 Social

3 Million unique installs

24 Million images generated

153 Years combined time spent on app

The Solution

With the directive to take LEGO to the masses, we wanted to keep the process of LEGOfying a photo via an iPhone app as simple as possible. We spent a good deal of time exploring the best way to guide the user to their creation quickly and intuitively. We gained clarity by minimizing the feature set, allowing us to spend more time focusing on the image algorithm and animation. Simplifying the interface and relying on smart iconography helped the app gain an international audience.

The Results

Upon launch, LEGO Photo was the number four free app in the United States, number two in the United Kingdom and number one in Japan. In the month of January, it attracted over two million unique installs, and ranked as the 32nd most downloaded application in the United States. There is even a flickr group to share and discuss LEGOfied photos.

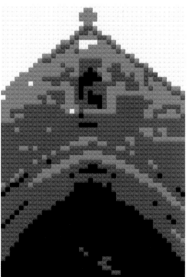

The World Park

Walk through
The Gates.
Again.

theworldpark.com

"One of the most inventive, alternative media
campaigns ever to invade Central Park."
Toni Fitzgerald, *Media Life* magazine

Client
NYC Parks & Recreation

Credits
Agency Magma
www.agencymagma.com

Awards
FWA Mobile; One Show;
Communication Arts

The Brief
Create awareness and engagement for New York's Central Park amongst younger, more wired visitors. Re-establish Central Park as one of America's first themed parks.

QR codes and image scanning were then just emerging in the US. However, they are poorly designed, so that's why the main goal for Magma was to create an iconic usage of a traditional QR code – we had a critical design challenge ahead of us. We believe that great design makes change easier. So, we knew that if we did a great job designing it, people would want to naturally interact with it.

Designing and introducing a new type of park signage to the public was also a design and usability challenge while having a low impact on the park both physically and visually.

The Challenge
Young consumers today primarily get their entertainment through the Internet, DVRs, and game consoles. Could we use the technology they use the most today, their mobile phone, as a key motivator to re-engage with the park? With the advent of the Nintendo Wii and smartphones, we had the opportunity to create an idea that motivated young people to come out and explore the world around them. The World Park is both innovative and social. It required them to earn park knowledge through their active, physical interaction with Central Park – meaning they had to actually walk the park and explore it in the real world to earn content in the virtual world.

The biggest challenge was creating, organizing, sourcing, licensing, packaging, designing, and writing all the creative content. Lastly, we culled it down and curated an outdoor mobile museum.

04 Social

The Solution

We turned the park into an interactive board game. An awareness campaign invited New Yorkers to an Arbor Day weekend event and also allowed them to unlock park content with their mobile phones through the advertising.

On the day of the event users could scan custom-designed QR codes resembling digital trees called PARKODES. Each PARKODE was color coded across four categories: pop culture, science & geology, art & music, and history. Each code was unique to its exact location in the park and delivered rich, relevant content from famous movie scenes, views from the 1800s, art, photography, streaming music, and the unknown facts about Central Park.

We wanted to educate park visitors while entertaining them, and we did so by connecting actual park locations to famous moments and facts in time.

The Results

The World Park event opened to the public on Arbor Day weekend 2010. More than 1,800 visitors used their mobile phones to interact with the park. The event gave New York City's Central Park a voice, a new medium to speak through and created a new way for tourists to interact with this iconic landmark. Overall feedback was very positive while most attendees were younger and played in groups.

The World Park wasn't a one-time event, it's a product, a piece of intellectual property. It was an event created to give Central Park exposure. With over 60 pieces of unique insights into the park's history the content isn't going to grow old any time soon. The next World Park event was scheduled for early 2011; the goal for our next event is to enhance the product and create a corporately funded event opportunity and revenue generator for the park.

1,500 Plus participants

120 Plus mobile interfaces

60 Plus PARKODE signs

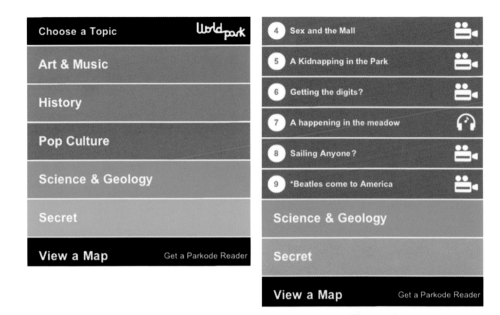

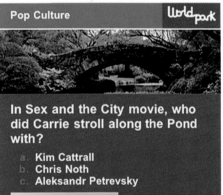

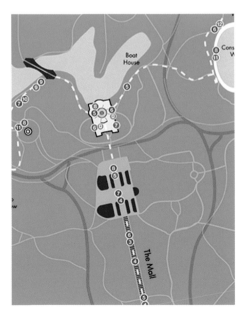

iHobo

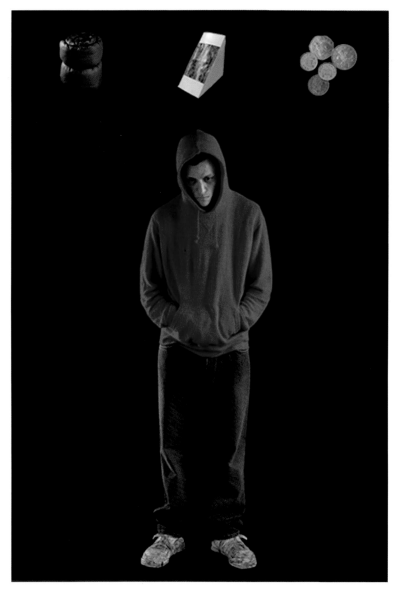

"I found the iHobo application effective but also disturbing in highlighting the realities of homelessness for young people. I admire the empathy that the app succeeded in drawing from the user."
iHobo user

The cold is getting to him. He's starting to feel really weak.

The Brief
Depaul UK is a charity that aims to keep young people off the streets. With an ageing donor base (average age 65+) it needed more awareness amongst – and donations from – affluent, socially-conscious 30-somethings in order to continue its valuable work. All previous campaigns had struggled to reach out to this audience.

The Challenge
From our conversations with the audience we knew if we could get them to really empathise with the plight of young homeless people we could get them to donate. There was one big problem though… they're very good at zoning out charity messages.

Many charity and public sector campaigns rely on one-hit shock tactics to get attention, but an audience that could happily walk past a real homeless person in the street would need to spend more time understanding the issue to feel empathy. We homed in on one big insight – whilst they find it easy to ignore a homeless person on the street, they find it impossible to ignore a beep on their phone. We decided to use their relationship with their phones to give them a longer-lasting, deeper understanding of life on the streets.

Client
Depaul UK

Credits
Publicis London
www.publicis.co.uk
Creative North
www.creativenorth.co.uk

Awards
FWA Mobile; Golden Drum Awards; Blades Awards

04 Social

He hasn't washed in a while. He's starting to feel really dirty and ashamed.

The Solution

We put a young homeless person on their iPhone. For three days the fate of iHobo was in their hands. Treated well, iHobo thrived. Left un-cared for, his inevitable decline into hard drugs played out on their screen (the shocking reality for three out of four young people who become homeless). After living with – and building a sense of responsibility for – him for three days and nights, the app prompted users to donate with one click.

The Results

Whether you judge this on awareness, or donations received, iHobo was an extraordinarily effective way of reaching out to a previously difficult to reach audience. iHobo was number one on the free app download chart a week after launch and to date it has been downloaded 502,600 times by 330,000 people. With a small production budget and absolutely no media spend, iHobo engaged people in a richly immersive way that would have cost at least £1.2 million in more conventional digital channels. Long-term ROI has yet to be calculated but is likely to be in the region of 6:1. Depaul UK gained unprecedented media coverage in the national, international and sector press.

502,600 Downloads

22,000 YouTube views

10,000 GBP money raised

4,000 New donors

Movie Mode
Mobile App

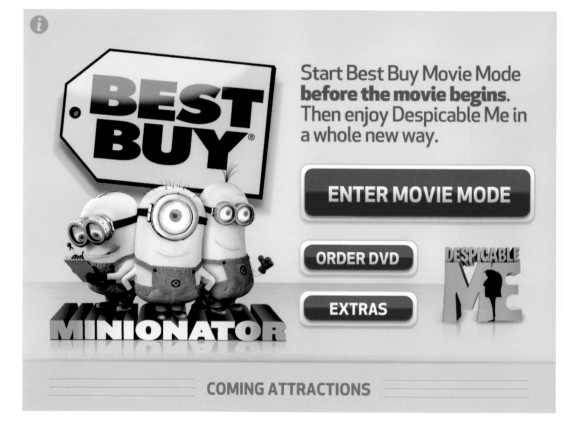

Start Best Buy Movie Mode **before the movie begins.** Then enjoy Despicable Me in a whole new way.

ENTER MOVIE MODE

ORDER DVD

EXTRAS

COMING ATTRACTIONS

"The app is a great example of how a brand and entertainment property can use the 'third screen' to connect with consumers while social viewing – which refers to individuals accessing their social networks or related content while viewing a TV program or film."
Paloma Vazquez, PSFK

The Brief
Our brief was simple. Develop a campaign that promoted the partnership between our client, Best Buy, and Universal Studios' feature animation, *Despicable Me*.

The Challenge
Typically movie tie-ins involve producing some ads, a special promotion or a limited time offer. For us the question was, how do we talk about this partnership in a way that does two things. One, demonstrates Best Buy's passion for innovating the way people use technology to enhance entertainment and two, does it in a way that'll generate some added talk value on their behalf.

Client
Best Buy

Credits
Crispin Porter + Bogusky
www.cpbgroup.com

Awards
FWA Mobile; Creativity Online

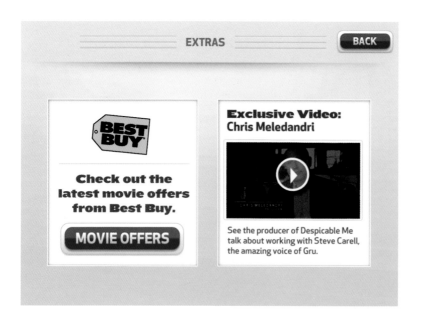

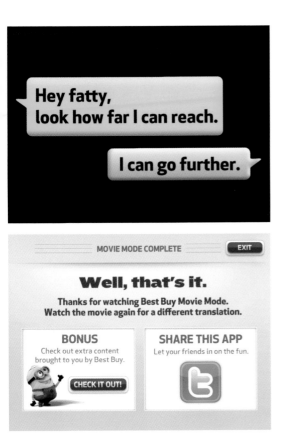

The Solution

As we watched the movie, we noticed a star role played by these little, adorable, yellow characters called Minions. While they were in nearly every scene, their language was completely indiscernible. They just spoke funny gibberish. And that's when it hit us. How can Best Buy enhance the way people experience *Despicable Me* on the big screen? By using the little screen. So we created the patented technology called Best Buy Movie Mode – the first ever mobile platform designed to interact in real time with movies on the big screen. For its debut, *Despicable Me* audience members didn't have to silence their phones and put them away. Instead, they kept them out and used Movie Mode to magically translate the gibberish spoken by the Minion characters into plain English, right on their mobile device.

The Results

Movie Mode was launched with a goal of 100,000 downloads. As of August 2010, Movie Mode has been downloaded over 343,531 times. The app was ranked as high as number six out of all entertainment apps and ranked #41 among free applications in the Apple App Store. Because of its revolutionary nature, the application and its experience have been the topic of tech blogs, news outlets and big entertainment publications alike. But perhaps most important is that Movie Mode is now being seen as a legitimate platform and a tool for movie studios to use to create additional layers of entertainment within their films. Right now, there are other Movie Mode experiences in the works for future feature films. As well as additional functionality planned for the at-home experience with the *Despicable Me* DVD.

100,000
Download goal

343,531
Actual
downloads

Opinionaided

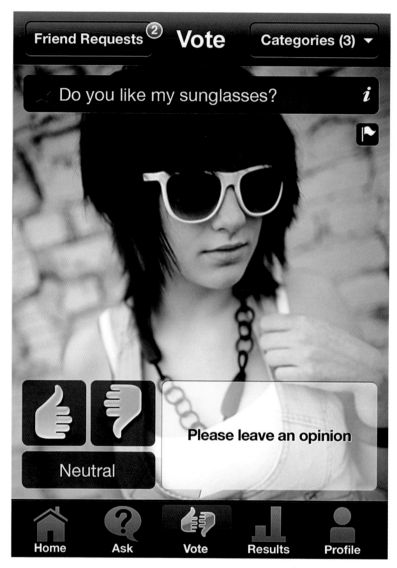

"Awesome and addictive app. Pretty good new features. Twitter support is cool."
Pixelinterrupt, App Store user review

The Brief

In the past, if you had a question, you would go to a search engine and have to sift through hundreds of pages of results, and after a while if you were lucky, you might have found something relevant. How long did that take? But what if you wanted an opinion? What if you're in a store and don't have all day? A search engine isn't going to be much help. You could ask your family or friends, but you're in a rush and need to know now. Opinionaided is a tool for helping people get relevant advice and opinions in real time. Ask a question, upload a picture, and instantly begin receiving advice and opinions from family, friends, and people all over the world.

The Challenge

If you're looking for facts you use a search engine. Google has that covered. But what if you're looking for an opinion? And more importantly, what if you're looking for an opinion relevant to you and your situation? Google can't help you here. Your friends and family can, but they can be biased sometimes, and they're not always around to help. Maybe you're at a store trying on a new shirt or you're at a restaurant and you want to know which dessert to choose – how can you get advice and opinions that are relevant to you immediately?

Client
Kurani Interactive

Credits
Kurani Interactive
www.kurani.com

Awards
FWA Mobile

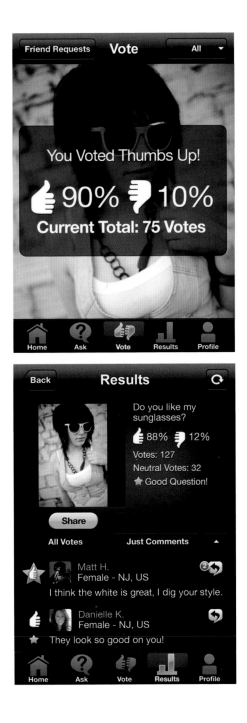

04 Social

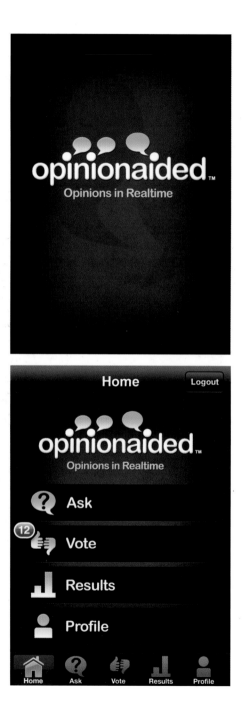

The Solution

Ask a question and receive real opinions from real people right when you need them. It's really that simple. Make better, more informed decisions based on what people think and how it relates to you. Connect with your friends and family, as well as unbiased strangers from around the world. Even make new friends based on what you have in common, not where you're from or who else you know. Opinionaided isn't a one-way street, either. We all have opinions and we all love to share them. Be helped; share your opinion, and help someone else.

The Results

In a span of six months, with over 120,000 questions and ten million votes, Opinionaided has proven that people are hungry not just for information, but also for opinions and advice, and specifically how they relate to them. Now with over 20,000 people using the service for an average of one hour a week, as the Opinionaided community expands and grows, the ability to receive relevant advice and opinions continues to improve.

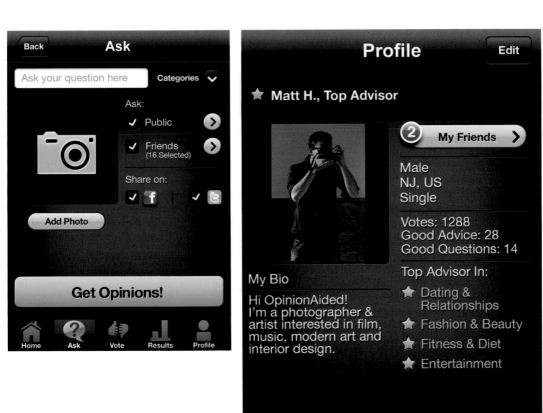

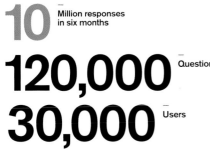

10 — Million responses in six months

120,000 — Questions

30,000 — Users

Alicia Keys OpenMic

"Rain did an unbelievable job for us. These apps
are more than we anticipated and ahead of their
time even in this age of technology. The fans
and the artists are having a blast with them."
Sean Rosenberg, Sony Music Executive

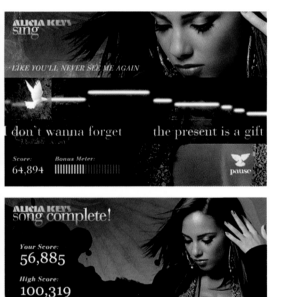

Client
Sony Music Entertainment

Credits
Rain
www.mediarain.com

Awards
FWA Mobile; Applied Arts
Magazine Interactive Award;
Davey Award; W3 Award; Summit
International Leader Award

The Brief
Sony Music Entertainment is a global record company representing music from every genre. Alicia Keys is a Grammy award-winning singer-songwriter who has sold over 35 million albums worldwide. With the success of Alicia's album sales, Sony wanted to push additional touch points between Alicia and her fans. With the continually increasing sales of Android and iPhones, mobile was by far the best platform to use for the project. The application would combine Alicia's music with an entertaining and interactive game, allowing users to play at any time and at any place.

The Challenge
Rain partnered with Sony Music Entertainment to create a mobile application that would establish a new point of contact between fans and Alicia. Promoting consumer interaction with the artist as well as other fans was a top priority for the project. Although pitch detection software did exist, the options available at the time did not match our needs. Rain was faced with the challenge of developing a new real-time pitch detection library capable of running on the iOS platform. Promoting Alicia Keys through peer publicity using social networking tools was an important aspect to include in the development of the app as well.

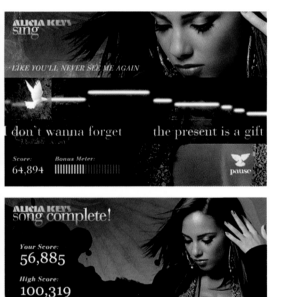

04 Social

The Solution
Fans were engaged with Alicia's music through a pitch-detection engine Rain developed that allows fans to sing along with their favorite songs and receive a score based on how accurately they match Alicia's vocals. To promote peer publicity the app is OpenFeint enabled so users can share their scores with the world. New achievements are unlocked in OpenFeint as the user's score improves. The scores are posted to a worldwide leaderboard, giving fans a revolutionary way to interact with Alicia Keys and with each other. The app also links to the Alicia Keys mobile site, blog, as well as to ticket purchasing for her tour.

The Results
After four months and 242,624 lines of melodic code, Alicia Keys OpenMic launched in the iTunes store. Since then the app has been very well received by her fans, receiving positive reviews and ratings. There are currently two free songs available with the app, with several more available for purchase. Sony Music Entertainment was thrilled with the Alicia Keys finished product, as well as the other OpenMic apps created by Rain. The Alicia Keys OpenMic app is Rain's most decorated project to date, receiving several industry awards since its launch.

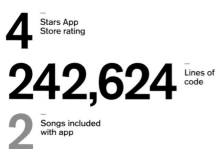

4 — Stars App Store rating

242,624 — Lines of code

2 — Songs included with app

adidas Urban Art Guide

"It's about giving information which you could never find out that quickly – especially when you are new in this city. That's what street art is about, to take the art to the people, to share it with them."
<u>NOMAD</u>, Artist

The Brief

On the occasion of its 60th anniversary in 2009 adidas Originals looked for an innovative way to celebrate the event according to the claim "Celebrate Originality".

While it was – of course – a goal to create media attention in relevant blogs and magazines, the main goal was to come up with a medium that met the target group of young, urban people "eye-to-eye". Starting with a local in Berlin the idea was additionally briefed to be adaptable for further roll-outs. Furthermore the idea had to be designed to be sustainable on the one hand but also give the opportunity for a corresponding event or celebration on the other hand.

The Challenge

When creating an idea for a target group that's young, cool and always on the move the biggest challenge might just be to be authentic and credible – staying true to the brand while hitting it off with the future users. In the case of adidas Originals the link between young, urban street culture and the brand's history was by no means far-fetched but still this relationship had to be emphasised in a way that neither ridiculed the brand nor became too commercial. Therefore a platform was needed to showcase this relationship in an authentic way without offending the target group in any way.

04 Social

Client
adidas AG

Credits
Neuland + Herzer GmbH
www.neuland-herzer.com

Awards
FWA Mobile; OnlineStar 2009
Gold Award; Global Mobile Awards
GSMA 2010 Nomination

The Solution

We developed the world's first mobile urban art guide for the iPhone and Symbian devices which was additionally adidas' first app worldwide. The adidas Urban Art Guide is an interactive app that lets its users discover their own surroundings or discover suggested walks with a focus on street art. Thus the adidas Urban Art Guide combines cutting-edge technology with a part of youth culture that began in the '70s and is still vivid and steadily evolving all over the world. Users of the Urban Art Guide can also be an active part of the app by rating, commenting on or recommending the pieces as well as discovering new artworks and uploading them.

The Results

The positive reactions to the launch of the adidas Urban Art Guide in Berlin meant that adidas expanded the Urban Art Guide to Hamburg in 2010 and is considering further international cities while continuing the work in Berlin and Hamburg. We have created a platform that brings more attention to street art and the artists. As a comprehensive app it not only addresses street art aficionados but also globetrotters who seek a new perspective and new angles on the city. Within the launch events for Berlin and Hamburg internationally renowned artists like NOMAD, Kami & Sasu or Milk created especially designed pieces thus actively contributing to the city's street art landscape – one piece at a time.

200,000 Downloads

82 Country downloads

17 Million online contacts

Videojug

"Wicked app, learned how to beatbox
while humming tonight and gonna
learn how to dance like Justin
Timberlake tomorrow."
Rustypegs, User

The Brief

Videojug is the original UK pioneer of "How-To" online video. Its aim is to help people to "get good at life".

Through its huge, ever-expanding library of professional, high-quality, instructional online video content, Videojug enables people everywhere to access valuable knowledge and advice on every subject imaginable – from cooking and health to technology and dating, in a straightforward easy-to-follow tutorial format.

People want to be able to access the content they want wherever they are.

The objective for the apps was to allow people to instantly access Videojug's videos on the go whenever and wherever a problem or a need to know how to do something presents itself. The aim was to create the handheld guide to helping you "Get Good at Life".

The Challenge

Our experience as the original "how-to" video website told us that time-poor people are searching the Internet for answers to their questions every day; and now with the advent of mobile devices they expect to be able to access the content they want wherever and whenever they want to.

Our challenge was to leverage our web publishing knowledge, taking it to the mobile platform to offer all the benefits of categorised short-form video in a mobile-optimised format and interface.

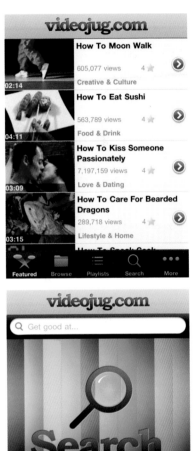

Client
Videojug

Credits
M&C Saatchi Mobile
www.mcsaatchimobile.com

Awards
FWA Mobile; App Store
"Staff Picks"

04 Social

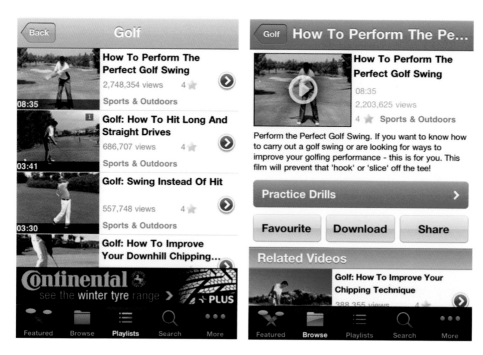

200,000 Downloads

150,000 Page views per day

68 Countries: in "What's Hot"

The Solution

Videojug's apps for the iPhone and iPad allow people to instantly access Videojug's huge library of professional, tutorial style how-to videos on the go, whenever and wherever they want to.

Unique features:
• Re-designed browser and category menu interface for iPhone and iPad to give you the film you want quickly and easily whilst on the go.
• Shake function – users can widen their knowledge and life skills just for fun and learn to do something new every day by shaking their iPhone/iPod Touch to instantly access one of Videojug's pearls of wisdom.
• Playlists – a catalogue of exclusive playlists covering popular topics including food, golf, dating, First Aid and DIY.
• Share function – users can share their favourite films with their friends via Facebook and Twitter.
• Download function: users can download films directly to their device to watch later.

The apps were created with a genuine purpose and user need in mind. Simple to use, providing fantastic depth and quality of content; a practical tool with longevity that enhances people's everyday lives.

The Results

The Videojug apps have netted more than 200,000 downloads since launch and generated over 150,000 page-views per day globally. In the App Store they have been ranked in the top five free Lifestyle apps – both in the US and UK – as well as being included in "Staff Picks", "New and Noteworthy" in the Lifestyle category and "What's Hot" across 68 countries globally. On December 9, 2010 the apps were nominated Mobile of the Day by FWA.

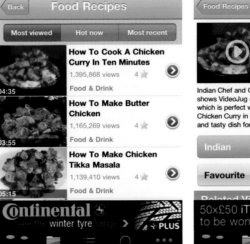

The Accidental News Explorer

"It's a rare application that changes the way you view the world. ANE is one of those applications."
Waxpraxis, US App Store user review

Client
Brendan Dawes

Publisher
CreativeApplications.Net

Credits
Design/Concept/Code
Brendan Dawes
www.brendandawes.com

Powered by
Daylife.com

Awards
FWA Mobile; New and Notable
in the US App Store

The Brief
I'm a big fan of chance encounters; randomly bumping into things can often lead to surprising and wonderful discoveries. But in order to bump into the unknown you have to be in an environment where chance is actively encouraged. Following on from my DoodleBuzz (www.doodlebuzz.com) project – a piece of work that was created to celebrate these opportunities for chance through a chaotic yet charming interface – I wanted to bring the same experience to the iPad so I could have the ability for serendipity away from the desktop not to mention make use of the multi-touch interface. It was a purely selfish brief – make something lovely just for me that would be worthy enough to sit on my iPad.

The Challenge
The thing is I'd never made an iPad app before. I'd never made an iPhone app either for that matter. On top of that DoodleBuzz is a crazy interface; there was certainly nothing standard about it. Forget using standard UI controls. Not only did I have to learn how to code in Objective C but I then had to figure out how to take what I'd made in Flash and port it over to this new language which seemed to be some form of geek Voodoo.

04 Social

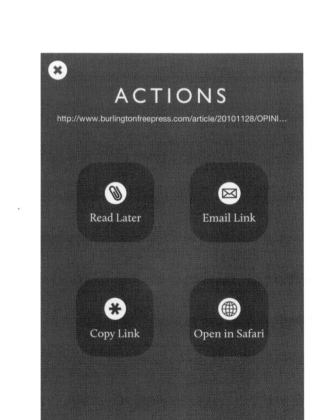

The Solution

So I did what I always do when presented with what seems like a huge task. I broke it down into its smallest parts and built small things gradually, which would then lead to a bigger whole. Early on I decided to work out how to get all the data in – via daylife.com – through a very simple iPhone app; I could then take that and build on the same code for the full-on iPad version. After weeks of reading three iPhone books, Googling like crazy and a continuing love-hate relationship with Objective C a funny thing happened: on the way to making DoodleBuzz for iPad I realised that I had a pretty nifty little iPhone app. Searching for news led to related subjects which led to more stories and took you down a rabbit hole filled with chance encounters. So I put the idea of DoodleBuzz for the iPad to one side and began to work on what I now called The Accidental News Explorer.

The Results

I made the app for myself. The early mornings crafting every pixel and every transition were made for an audience of one, which was me. If other people liked it then that was a bonus, but I was under no illusion that this was an app for everyone. But when I get feedback from people who tell me how much they love it, how they even missed their stop at the train station because they were so engrossed with the Accidental News Explorer, and the things they'd discovered purely by accident, then that makes me smile. This app is just my little way of helping people discover the delight of chance encounters.

238
Four-letter expletives used whilst developing

9
Features removed to make a better product

17
Amount of times I found the answers on Google

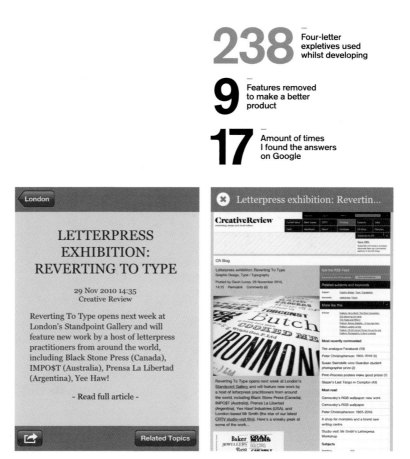

The Next Move
by UrbanDaddy

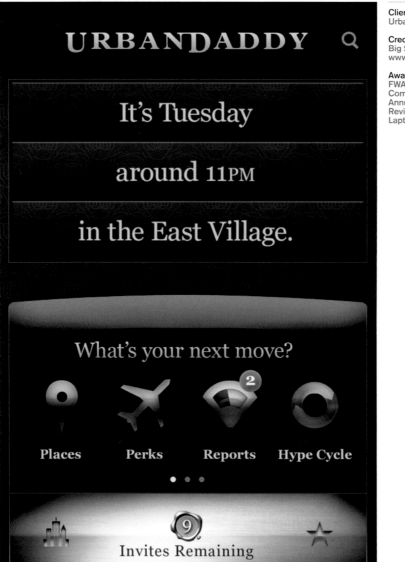

Client
UrbanDaddy

Credits
Big Spaceship
www.bigspaceship.com

Awards
FWA Mobile; Webby Awards;
Communication Arts Interactive
Annual; Hive Awards; Creative
Review Annual; Mobi Awards;
Laptop Magazine; OMMA; W3

"UrbanDaddy's The Next Move is an iPhone app
empowered with the one quality that even the
most powerful robotic search engine probably
won't even learn: really good taste."
Esquire.com

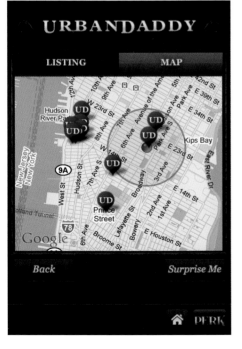

The Brief

Our goal from the beginning was to build an app that allowed people to carry an UrbanDaddy editor in their pocket at all times. We set out to take the core principles of our brand – curation, a sense of adventure, an appreciation of luxury, a love for the local scene, and a fair helping of wit – and deliver that to our readers on the go.

We've built up a trust with our readers over the years. So we set out to create a unique, beautiful product that maintains that trust, delivers only the best of the best in each city, and also maps to how our readers use the city while they're out and about. Something that answers the question "where should we go next?" in the most elegant, logical way possible.

The Challenge

When we started to think about what UrbanDaddy would look like on a mobile device, we wanted to make sure we represented the spirit and the essence of our brand while still creating new value for the platform. We didn't want to simply replicate our website and email experiences on a phone. We didn't see much value in that exercise and we knew that the editorial content and city knowledge that we had built up over the years had value to readers on the go.

So the challenge became how to deliver our editorial opinion on the city to our audience in a way that made sense on the move. And to do it with a little panache.

1 Million plus
downloads

9 Cities

16 Minutes plus average
time on app

270,000 People looking
for cougars

190,000 People looking
for celebrities

1 Apple print campaign
featuring app

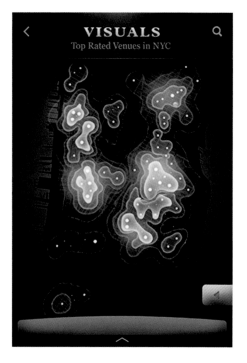

The Solution
The idea started with a few questions: Where
are you? What day is it? What time is it? Who
are you with? What do you want? Using the
power of the iPhone, we were able to answer
the first three questions for the user.

From there, the user can home in on
the perfect place for their needs at that very
moment. Oysters at 3am with your girlfriend
on a Tuesday in Gramercy Park in New York?
Dancing with your mistress in the mission in
San Francisco at 8pm on a Sunday? A sudden
urge for champagne with colleagues at 4pm on
a Friday in the Back Bay of Boston? These are
the questions that the UrbanDaddy iPhone app
was born to answer. Targeted results – only
the best of the best places in your city –
categorized the way you already think, and
delivered by distance in list and map form.

The Results
The app was a runaway success, bringing
in a new and growing wave of UrbanDaddy
members. It's been downloaded over one
million times and Apple featured The Next Move
in a national print campaign that ran for months
on the back of every major national magazine.
The app also made a brief but shining cameo in
an Apple TV commercial. But the biggest result
for us is probably the loyal following we have
been able to garner through The Next Move.
We've committed to pushing the boundaries of
what's technically possible on the iPhone and
creating beautiful features for our members
going forward.

Macmillan Coffee Finder

"We wanted to come up with an interesting and fun tool for users which also helps Macmillan spread their message whilst making it easy for people to donate and take part in the World's Biggest Coffee Morning."
Noel Lyons, Director, KentLyons

The Brief
Every year, Macmillan organises the World's Biggest Coffee Morning, an event involving millions of people, getting together to have fun, make friends and raise money for cancer care and support.

We wanted everyone to get involved – all year round. This iPhone app is designed to help people find coffee shops to meet up with friends and to spread the campaign message for Macmillan.

The Challenge
Starting out with over 900 cafés and coffee shops reviewed and rated around the UK, we've input what we think are the best coffee shops. Users can also rate cafés by giving them a mug rating (from one to five), upload pictures of the café, the pastries, the view etc. and comment on them too. Using our clever filters you can get more specific – find coffee shops with gallery spaces, Wi-Fi, toilets, al fresco seating and organic produce.

04 Social

Client
Macmillan Cancer Support

Credits
KentLyons
www.kentlyons.com

The Solution

This great, free app from Macmillan Cancer Support helps you find the best cup of coffee near you. Using our map of the UK, you can easily see where the cafes are.

Better yet, using our amazing augmented reality technology, you can see a floating green mug representing the direction to the café – all you have to do is follow it to find a great mocha chocca latte. This Macmillan app renders live text using fonts that are not native to the iPhone.

The Results

If the app helped you, it suggests donating the price of a cup of coffee with its in-built donation button, thereby raising even more charity funds. Plus, you can sign up to host your own coffee morning, all within the app. It's thinking like this that allows Macmillan to help people have a good day. You can find high-quality, reviewed and reliable cafés and coffee shops for free.

900
Cafés and coffee shops reviewed

7.8
Million GBP raised, partly by text donations

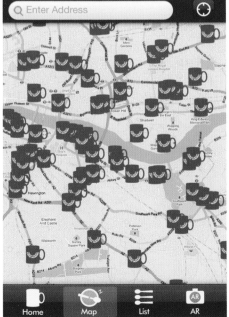

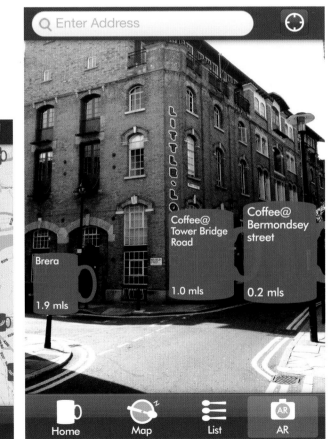

Instagram

"Over the last two months I've watched friends
become slightly obsessed with Instagram. I have
too. We constantly check to see new photos from
friends and engage in chatty conversations with
strangers around the globe about their images."
Nick Bilton, *The New York Times*

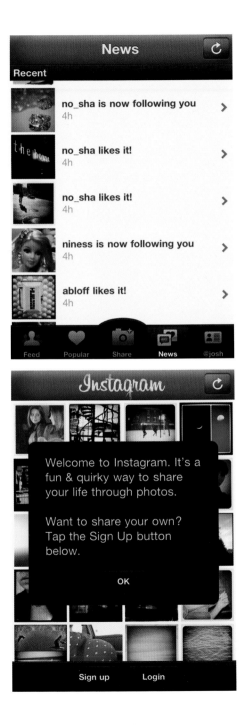

The Brief

When we were kids we loved playing around with cameras – we loved how all the old Polaroid cameras marketed themselves as "instant" (something we take for granted today). We also felt that the snapshots people were taking were kind of like telegrams in that they got sent over the wire to others – so we figured why not combine the two?

Instagram came from that inspiration – could we make sharing your life as instant and magic as those first Polaroid pictures must have felt?

The Challenge

When we sat down in the summer of 2010 to start designing our product, we looked at digital photos and realized very few exciting things had happened in the last five years. In that time, mobile cameras had improved tremendously, but the photos they produced lacked mood and tone. On mobile devices, photos can take a long time to upload, and viewing them is slow.

Aside from those technology problems, it can be a pain to share with all the friends you care about, across multiple social networks. We set out to build a mobile app that makes sharing mobile photos fast, simple, and beautiful.

Client
Burbn, Inc

Credits
Instagram
http://instagr.am

Awards
App Store Featured;
#1 Free Photo app

04 Social

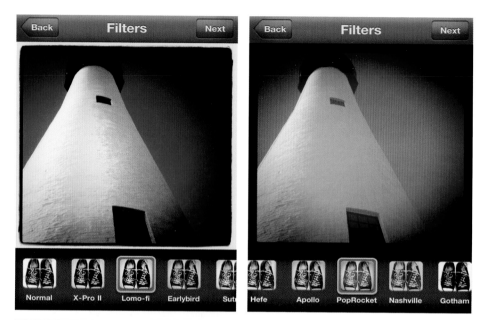

1 Million registered users
in three months

3 Photo uploads
per second

650,000 Website visits
per day

The Solution
We built a fast, beautiful, and fun way to share your life with friends through a series of pictures. We gave users the ability to add mood and tone to their photos by using 11 one-click filters in the app that instantly transform photos into Lomo-Fi shots or vintage, sepia-toned images, among other effects. We worked to make the uploading, sharing, and viewing experiences as smooth and speedy as possible for all users, whether you're using an older iPhone or the latest version. And we made it super-simple to share photos not only with your followers in the Instagram community, but with your friends on Facebook, Twitter, Flickr, and Tumblr, all with a tap of a switch.

The Results
We never expected the overwhelming response we received. We went from literally a handful of users to the number one free photography app in a matter of hours after our launch in the App Store on October 6, 2010. Over 100,000 users registered for Instagram in our first week. We were the featured App of the Week in the App Store three weeks after launch. In just three months, over one million people were using Instagram as a way to communicate with their friends through photos, including major organizations such as National Public Radio (@npr), Starbucks (@starbucks) and NBC News (@nbcnews). We've also seen huge adoption internationally; as a result, we've translated the app into seven different languages.

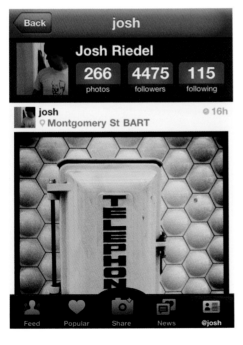

04 Social

Utilities

Introduction by
Remon Tijssen, Adobe Systems, Inc

05

Traditional desktop utility software is typically designed to help with computer-related tasks or processes: reading computer-related data like processor speed and disk space, optimizing the computer's performance by putting the bits and bytes in a better structured order or even worse, dealing with problems like viruses.

Being little helpers that are often needed, but rarely desirable, they look up to their big brothers and sisters called applications that allow their users to create stuff, consume content and play games, software that brings a deeper satisfaction or delight when using it. The emotional potential is greater. In contrast, when designed and/or engineered wrongly, greater frustration or disappointment can occur.

"What we called widgets on the desktop, could now be called a utility application on a mobile device. A simple weather viewer, a calculator, a stock exchange viewer, etc."

Naturally, among the first utility applications for mobile devices, these little helper applications that are more related to the well-being of the machine than the person using it, when converted to their mobile context, they were still useful and at times painfully needed.

The category was quickly broadened by literally taking the small technical helper application meaning more loosely. Now applications with, for instance, little interaction, displaying or measuring a broader type of data, are called utility applications as well. What we called widgets on the desktop, could now be called a utility application on a mobile device. A simple weather viewer, a calculator, a stock exchange viewer, etc.

At the time of writing this piece, looking at what device and device-operating system manufacturers have listed under utility applications, it becomes clear that change is upon us.

Android doesn't have a utility category, but if we look at the category that is traditionally closest to utilities, the tools category lists a voice translator, a compass & level, an uninstaller (really?? I'm afraid so…), and a lot of different flash-lights.

Windows mobile is more true to the traditional meaning of a utility app, a broad variety of little applications that measure and display technical data and try to keep your mobile device in good shape: apps that back up, take notes and reset your device to name a few.

Among the top paid utility categorised applications for iOS were those that scan bar codes, improve photos, browse the web, ring an alarm, broadcast police radio channels, and more flash-lights.

So, what's going on? Did the utility category for this new type of utility applications become an extra bucket for attention, are they left-over apps, apps that don't fit well in more clearly defined categories like games, music and news?

Going back to what's desirable, the big desktop applications might have initially and by concept set themselves up for greater use and satisfaction for a user, but if created wrongly they outgrow themselves trying to please each and every one of their users.

"The mobile space is a venue for liberating traditional approaches to application development that are consciously embraced, but also to generate new results without always realizing it because of technical and production limitations."

The meaning of what an app is has changed pretty dramatically as you may have noticed. Even a piece of content can be wrapped as an application. Applications are more focused and dedicated. Five well-chosen effects that can be applied to your photos can be a great application that people would even pay for. It's not about the numbers, it's about the best possible way of getting things done and being directed by the application which is best for you. At this point in time, we just had the first selection of applications that were able to take just a few features and employ them really well versus taking as many as you can and not really caring that much about the user experience. The mobile space is a venue for liberating traditional approaches to application development that are consciously embraced, but also to generate new results without always realizing it because of technical and production limitations.

Utility software seems not to be the most exciting type of application, but when you think about it, and take the meaning of a utility app – to assist the user in the context of a larger task – we have the start of a new model for applications.

The next generation of applications can learn a lot from those little helper applications back in the days. The most important ingredient that they miss is connectivity with other apps.

It's going to be very interesting how more powerful and feature-rich applications, new or ported from existing desktop applications, are going to be designed for mobile devices. Are we going to bloat the screen again with visual UI and confuse people with inconsistent and confusing touch gestures to control all those tasks? Or are we going to see an expansion of contextualized and customized apps that focus on specific tasks where work-flows can happen across these apps because we set them up for it?

"It's going to be very interesting how more powerful and feature-rich applications are going to be designed for mobile devices."

What happens when two applications or application components are plugged into the same functional and UI framework to mix and match? Is this complicating the user experience or can we make it adaptive enough and part of the natural work-flows, so it actually enlightens the user experience? What if all the content that is viewed, extended, changed and created across these small applications and devices has access to all other content, constantly complementing and creating real-time logical work-flows for that user for that moment? We'll have to see how this is going to work or not in different contexts, but this connectivity between apps is going to be a potentially big thing in the next generations of mobile applications.

There are a lot of hurdles to take, and outdated rules can slow us down and even prevent us from getting to this ultimate situation, but this is an interesting step to make and something which is natural to people.

Back to the sort of utility app that originates from little desktop tools that help you monitor and maintain your computer. They are still part of mobile utility applications and they will probably be here for a while on mobile devices today, but the rules have clearly broadened. If this hyper-connected application world is going to accelerate, let's hope there are no typical monitoring utilities needed any more and it all just works.

Remon Tijssen
Adobe

Bio.
Remon Tijssen
Adobe

Remon Tijssen focuses on a broad spectrum of digital design: dynamic software, games, toys, motion graphics, and presentations where he constantly moves in and out of the high-tension zone between playfulness and functionality.

In 2000 he founded Fluid.nl in the Netherlands. Under the umbrella of Fluid he defines new approaches to interaction design and continually explores the wider possibilities of digital media.

Remon's passion to inject motion and play into his inventive interface solutions and his specific, world-renowned personal touch to dynamic behaviors, accompanied by a strong, solid visual language, have brought him international recognition and awards.

Since 2008 he has been a Senior Experience Design Lead for Adobe in San Francisco where he is working between the horizon and the near future, exploring new technologies, creating new experiences, and delivering projects that demonstrate Adobe's technologies and leadership. Recently he worked on paving the way for a new generation of mobile and touch applications with Adobe Ideas as the first to see the light of day.
—
www.adobe.com
www.fluid.nl

"If this hyper-connected application world is going to accelerate, let's hope there are no typical monitoring utilities needed any more and it all just works."

Adobe Ideas

"I've been sketching like a maniac using Adobe Ideas…
It's been a great tool so far and has dramatically
changed my sketching habits. Not to mention clients
are impressed and the sketching images produced
from Adobe Ideas are of great quality, good enough
to fold into client presentations in fact."
Jess Eddy, User Experience Designer,
New York, USA

The Brief

Tablets and other mobile devices have revolutionized traditional work-flows and experiences. Users are free to do much of their work from anywhere. They can read and send email, review documents and presentations, and join conference calls just about anywhere.

While creative professionals can do these things from mobile devices as well, these tasks are only tangentially related to a designer's core work. Tablets are poised to free designers as well by offering the ability to integrate mobile devices into key aspects of the creative work-flow, whether it's for content creation, sharing, iteration, or presentation. Adobe Ideas frees creative professionals by allowing them to create anywhere, any time.

The Challenge

Adobe wanted to create an application that would stimulate creative exploration and fit into professional creative work-flows, but it also had to be approachable and simple enough for anyone to pick it up and use it right away. In this case, Adobe was not looking at the tablet platform as a tool for creating finished artwork, rather as a portable digital sketchbook to supplement what artists and designers do with Moleskine or paper napkins. Since touchscreens do not typically compete against the precision and tactile nature of a pencil, the challenge was to find ways in which this medium could add unique value.

Client
Adobe Systems, Inc

Credits
Adobe
www.adobe.com
Image Credits
Crab and Rooster: Carol George
Obama photo: image courtesy of AP
Obama autograph: Sylvester Cann IV
Man at bar: Vitor Peres
Adobe Ideas Boy: The Fat Beicon
Lady sketch: Stefan Marjoram

Awards
FWA Mobile; Mobile Star Awards

05 Utilities

The Solution
Adobe Ideas provides simplicity, scalability, and compatibility with professional tools. Adobe Ideas has a very limited tool-set that makes it immediately approachable. However, the profound thing is that creative exploration is actually enhanced by the simplicity. Complementing this minimalist tool-set is the incredible ability to zoom in and out while sketching and always see cleanly rendered vector lines. Since creative professionals can zoom without seeing big pixels, an amazing amount of detail is possible. Compatibility with professional tools comes from the export file format – PDF. Ideas that are emailed from the app are vector-based PDFs that fit perfectly into work-flows using Adobe Photoshop, Adobe Illustrator and other pro creative tools.

The Results
Adobe Ideas has been well received by the users, press, bloggers, and reviewers. Almost daily, the Adobe Ideas team finds a positive comment on Twitter, Facebook or on a design site. Adobe's customers pleasantly surprised the Adobe Ideas team by showing various uses and work-flows and continuing to download Adobe Ideas. Even President Obama used Adobe Ideas to sign his autograph – this is the first known time that a US president has signed an iPad. Adobe Ideas is really something special. Adobe has an innovative team behind the technology, and they value creative professionals and intelligent user experiences. This application could be the beginning of more exciting things to come from Adobe.

Mr. President, please sign my iPad. The highest-profile user of Adobe Ideas known is US President Barack Obama. In late October 2010, less than a year after both the iPad and Adobe Ideas were launched, Sylvester Cann IV was at a rally at the University of Washington, in Seattle, Washington. He fired up the Adobe Ideas app on his iPad and asked the President to give him his autograph using the touchscreen. Obama said that the Secret Service was unsure but they soon gave the green light. Obama looked surprised, but used his finger to scribble on the iPad. Cann said, "This HAS to be the first time an iPad has received a Presidential autograph." Adobe is proud to be part of this known first.

51 Percent users
updating to v1.1
in 10 days

#1 Highest-profile
user: US President
Barack Obama

Staphylococcus Aureus — Transmitted: Eating Contaminated Foods & via Respiratory tract

Streptococcus — Transmitted via Respiratory tract

HIV — Symptoms flu-like

BACTERIA

Salmonella — Transmitted Eating Contaminated Foods

Schistosoma (Flatworm) — treatment: Single dose of Praziquantel

Pathogens

Candidiasis — Treatment: Antibiotics

WORMS

Cestoda (tapeworm) — Prevention: Keep hygienic & wash your hands

Hookworm — Treatment: Drugs such as Mebendazole

Chicken Pox — Symptoms Red spots, Itching

VIRUSES

Hepatitus — Symptoms headaches, fever, nausea

Histoplasmosis — Treatment: Antifungal medication

FUNGAL DISEASES

Athlete's Foot — Treatment: creams, good hygiene

BBC News

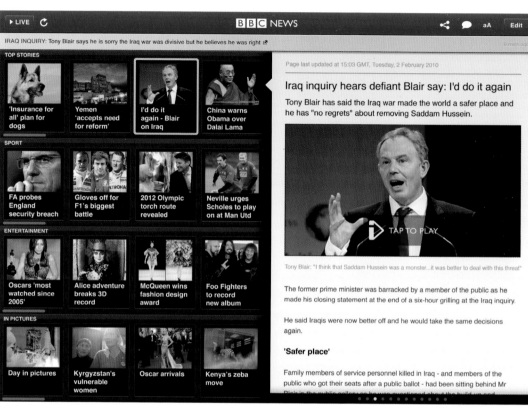

"If you were going to design the TV news broadcast
of the future, where viewers pick the stories and
watch them come to life, this is it."
USA Today

The Brief

BBC News was a latecomer to the iPhone App market. Two years into the iPhone App Store, Apple announced that it was launching a new device – the iPad. This was the perfect opportunity for the BBC to join the mobile app market with its own much anticipated News app. Our goal was to produce an app which differentiated BBC News from the competition by designing a fresh way to navigate and deliver rich media and text content, as well as to serve up the latest breaking news from the BBC in a friendly, intuitive and usable manner. The brief was broken down into two keywords – skim and dip.

The Challenge

When we started work on the BBC News iPad app we were, in a sense, designing blind as the iPad, as a device, hadn't been released yet and no one on the team had physically seen or played with an iPad. A huge amount of work needed to be done to ensure that the app was complete in time for the launch of the iPad. We knew the basic specs of the device and used that, combined with our previous iPhone App experience, to shape the future app. As we had no device to test our designs on we made use of print-outs to test the size and placement of elements on the screen. It was essential to design an app that was unique in its look and feel but which represented the BBC brand.

Client
BBC

Credits
The Noble Union
www.thenobleunion.com
Mobile IQ
www.mobileiq.com

Awards
FWA Mobile; #1 free app
in UK App Store; #1 news app
in 16 countries, Top 10 in 62

05 Utilities

3.5
Million downloads

722,000
Users per week

45
Percent users
using app
every day

70
Percent users
using app
every 3 days

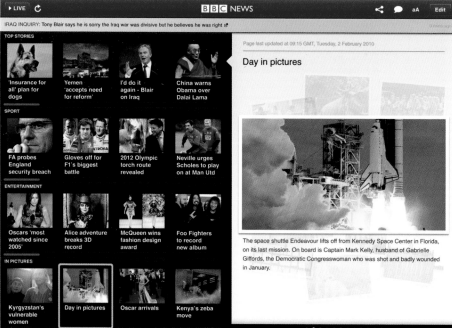

The Solution

Working with the keywords "skim and dip" we set out to devise a simple mechanism that would allow users to gain an overview of the news as well as the ability to view and switch between news stories quickly. We designed a navigation system based around scrolling carousels which would enable this "skim and dip" metaphor to become a reality as we could simultaneously show a multitude of news headlines. The carousels incorporated thumbnail images for each story making the overall effect of browsing news a visually attractive mosaic. Splitting the screen allowed us to instantly show the selected news story alongside the multitude of news headlines presented in the carousels.

The Results

The BBC News iPad and iPhone apps have become the top downloaded apps in many countries and been added to many must-download lists. The apps have set a new standard for news apps – scrolling carousels have since been replicated by many other apps. The number of news stories read by users, when compared to the BBC News website, also increased dramatically, possibly due to the ease of being able to tap stories on the home screen and have them appear instantly in the reading pane.

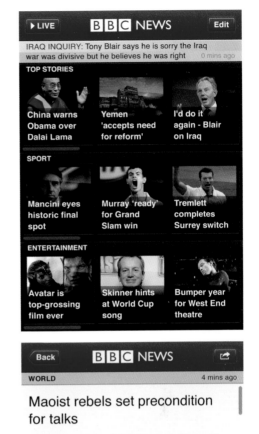

Met Office iPhone Weather App

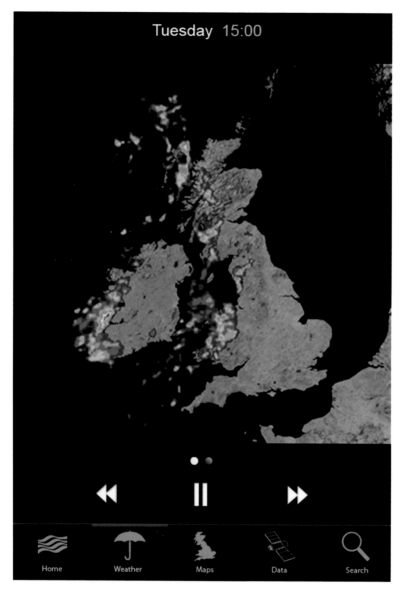

"This has been our first entry into mobile. The iPhone has proved to be a fantastic opportunity to understand the mobile market. Users can leave feedback and rate the service – which has been invaluable and helped us to direct and steer new services."
Mona Lukha, Project Manager, Met Office

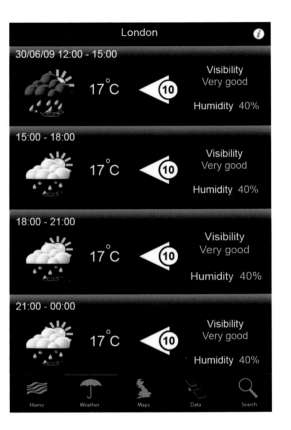

The Brief

If you were asked to name an innovative company, what would you say? Apple, Dyson, Google? One name that doesn't often come up is the Met Office, something that we're working hard to change. The initial idea for the iPhone app came out of a workshop held with Met Office staff in 2008. The workshop was designed to identify potential ideas for increasing the reach of the Met Office's Public Weather Service. Our main aim was to enhance the way the Met Office delivers weather products and services to the public, our customers and stakeholders.

The Challenge

We wanted to develop services to complement the various ways of getting the latest information from the Met Office – for example we already provide information through our website, RSS newsfeeds, Twitter, email alerts, and videos on YouTube. We identified several possible technical and promotional opportunities and investigated them further. At the time, the iPhone was new to the market and we recognised that it would be an important product in the future and that it would be beneficial to have an application in the Apple Store. iPhone users already have a built-in weather application on the device, but we wanted to develop an app that would provide much more detailed weather information.

Client
Met Office

Credits
Gorillabox
www.gorillabox.tv

Awards
Apple Featured App

05 Utilities

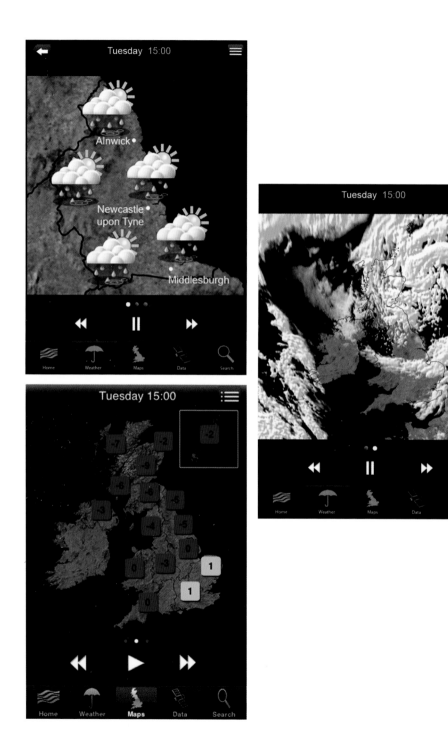

The Solution

In January 2010, we launched our free iPhone weather app – providing five-day forecasts, national severe weather warnings, daily national and regional weather maps and UK rainfall radar and satellite imagery. The app uses GPS to give users location-based weather. For the planning, managing and tracking of the iPhone weather app we worked with mobile developers Gorillabox. As well as the iPhone app, we've also developed content for other web-enabled phones. Designed with fewer graphics, they provide essential information that loads faster than a standard website.

The Results

Since its launch, our innovative iPhone weather app has become the top free weather app in the UK, with over 1.4 million downloads. The app was featured by Apple after just one week in the App Store. It now ranks number seven in the overall Top free iPhone apps list – a significant achievement, especially considering there are more than 50,000 free apps available.

With so much competition, new apps have to be good to stand a chance of becoming popular. There is also opportunity to introduce apps for specific groups such as surfers, sailors or mountain climbers. Where there are weather-related activities, there's an opportunity for an app in its own right.

1.4 Million downloads

21.5 Million visits

711,292 Visits per day

151.8 Million hits

MathBoard

"This is the best integer practice app or program
I have found in 34 years of teaching math."
Tim Seiber, Math Teacher

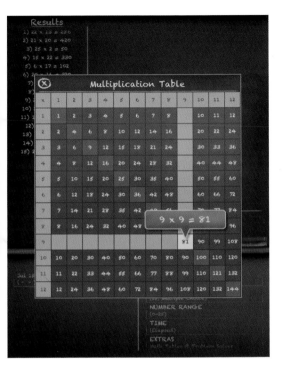

The Brief

MathBoard is an educational app designed to help children learn and master their math skills. As a long-time software developer and father of two, Paul Schmitt aspired to teach kids math in an enjoyable and rewarding manner. The aim was to help kids master basic skills such as simple addition and subtraction, as well as more difficult concepts like multiplication and division. Making this process engaging for today's school-age children would inspire them to practice and become more proficient in their knowledge of math. And so began the development of valuable apps such as FlashMath and FlowMath for the iPhone and iTouch. These fun apps allowed kids to practice their math any time and anywhere. With the launch of the iPad the ability to expand on these initial ideas became a reality. The potential to add more features, including greater number ranges, varying skill levels, and timed quizzes was now achievable. And so MathBoard was created.

The Challenge

Design an educational app aimed at teaching kids math. It seemed like a simple vision. However, it also needed to be visually appealing, enjoyable, and fun, so kids from Kindergarten through Elementary School would actually want to "play". In addition, it had to be educational and comprehensive so children would be able to learn and master their math fundamentals. Lastly, worthwhile so parents too would agree that this is truly a valuable math app for their child. All these were challenges faced by the developer. However, the greatest challenge was the "Problem Solver" feature. This is where the individual steps for solving addition, subtraction, multiplication and division problems would actually be shown. This step-by-step explanation is where MathBoard set itself apart. The child is now actually being taught how to do each specific problem, rather than just being told whether his or her answer is correct. With the mastery of each level, the complexity and therefore the learning can be increased. This is a vital part to making MathBoard an exceptional learning tool.

Client
PalaSoftware Inc

Credits
PalaSoftware Inc.
www.palasoftware.com

Awards
FWA Mobile; iPhone and Kids

05 Utilities

The Solution

Using a traditional chalkboard as the background, the nostalgic appeal for parents was instantaneous. Having the iPad be the new slate, with no more chalk dust or mess, made it even better. Using varying skill levels, number ranges, problem solvers, and time limits made the goal of teaching a wide range of children achievable. Adding multiple choice and fill-in-the-blank type quizzes, the ability to save data, and work out problems on the expandable "chalkboard" completed the vision of making it enjoyable. The reviewing, learning, and mastering of math skills any time and anywhere became an attainable goal. What children now see as a fun game to play, parents know is both beneficial and educational. This is the genius behind MathBoard.

The Results

The success of MathBoard has exceeded expectations. Since the first week of its introduction MathBoard has consistently been at the top of the educational software charts. In just six months, MathBoard has had over 100,000 app downloads. It is available worldwide, and has been translated into six languages. It has also been featured in the iPad Delicious TV ad, which is aired in multiple countries. To date MathBoard continues to be one of the top learning tools in all Apple App Stores worldwide.

6 Language versions

#1 In App Stores for Education: 43 countries

100,000 Plus downloads

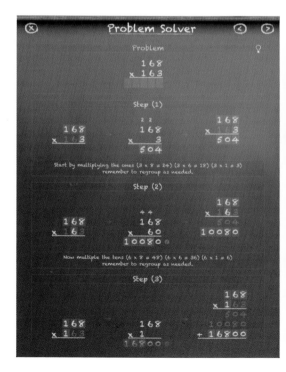

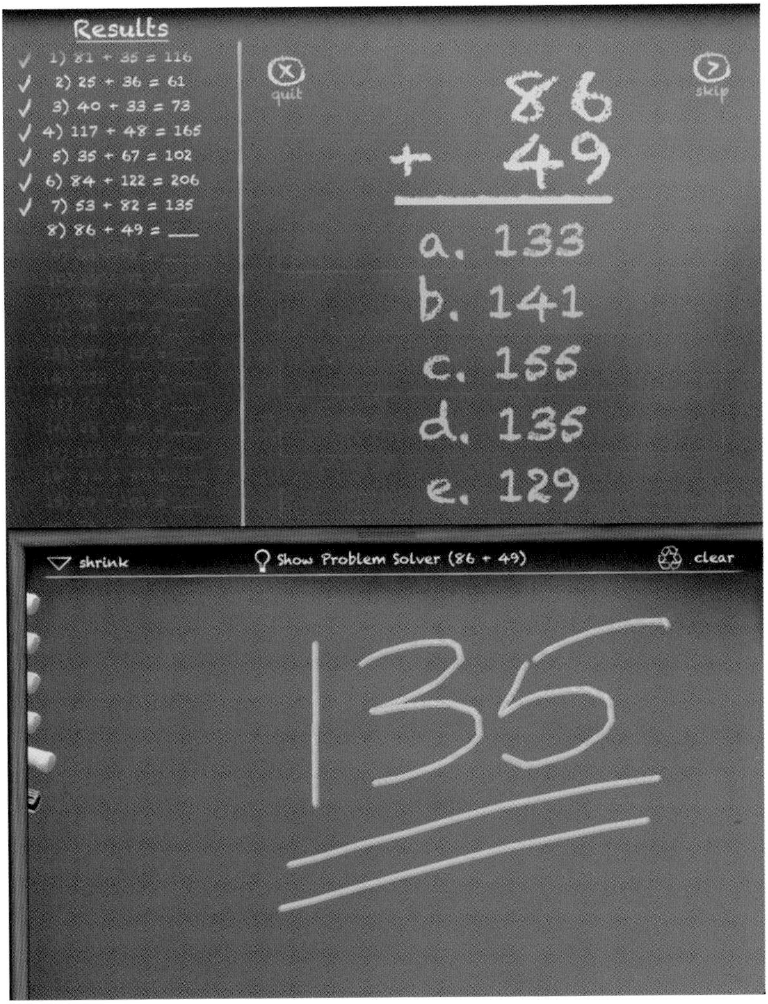

Results

✓ 1) 81 + 35 = 116
✓ 2) 25 + 36 = 61
✓ 3) 40 + 33 = 73
✓ 4) 117 + 48 = 165
✓ 5) 35 + 67 = 102
✓ 6) 84 + 122 = 206
✓ 7) 53 + 82 = 135
   8) 86 + 49 = ____

quit

skip

$$86 + 49$$

a. 133
b. 141
c. 155
d. 135
e. 129

▽ shrink   ♀ Show Problem Solver (86 + 49)   ♻ clear

135

SAS Survival Guide

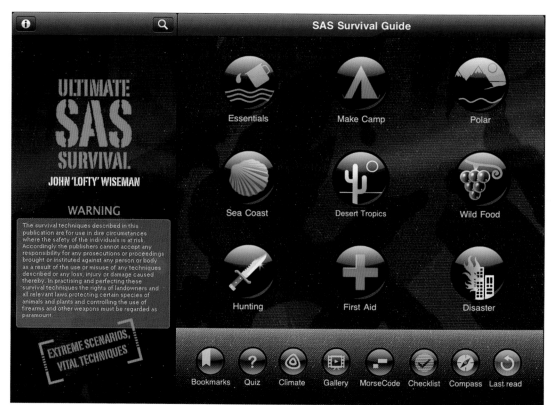

"What makes the guide come alive, though, are the other cool features: quizzes testing your survival knowledge; photo and video galleries that take you step-by-step through activities like building fires and setting up camps; a compass; and – neatest of all – the ability to type out and convert emergency messages to Morse code, using light flashes to signal other nearby adventurers."
Macworld

Client
Trellisys.net Pvt. Ltd
HarperCollins Publishers

Credits
Trellisys.net Pvt. Ltd
www.trellisys.net
Trueways Survival

Awards
FWA Mobile; App Store: iPad App
of the Week, Featured App,
New & Noteworthy, What's Hot

Table Of Cont...   **Hunting**

DANGEROUS CREATURES

The insects and other creatures shown here are not a major problem for survivors if sensible precautions are taken – but can easily become one if not treated with respect.

1 SCORPIONS are found in deserts, forests and jungles of tropical, subtropical and warm temperate areas, one kind living at 3600m

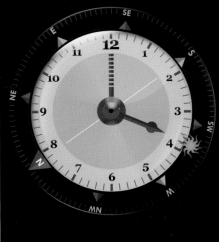

Back   **Sun Compass - Step4**

Step 4

Bisect the angle between the hour hand and the 12 mark. The compass needle now points North-South

The Brief

HarperCollins contacted us to discuss their plans for creating an app from the million-copy best-seller *SAS Survival Guide* book. The Brief was not simply to include the entire book content in a great readable format, but to add function and multimedia to transform the book into a lifestyle utility app.

The Challenge

The iPhone Development Team at Trellisys.net then got into action. We sat down in an attempt to define the challenge. An analysis of various other apps on the iPhone revealed that users not only demand depth of content but also stunning design coupled with readability and simple usability.

The Challenge then was to create an app with the book at its core and some practical utility apps packaged into an easy-to-use, interactive and immersive experience. Importantly, we had to decide on a content format that stays true to the two-column layout of the book interspaced with lots of illustrations, how-tos and photographs. The next challenge was to identify handy bolt-on micro apps that an outdoorsman would find useful.

05 Utilities

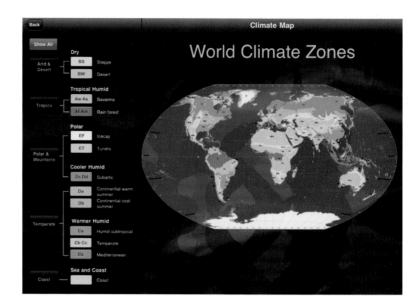

852 Five star App
Store reviews

1 Million
sessions

200,000 Downloads in
seven months

The Solution

We started off with a beautiful design built around army fatigues and colourful, easily identifiable icons that demonstrate ruggedness and adventure.

We decided to convert the entire book into a single-column layout so readers only have to scroll in a single direction. We built a more comprehensive "Table of Contents" and "Search" feature so users can jump directly to the content they were interested in. We also built a "Resume Reading" and a handy bookmarking feature.

The immersive experience was created by seamlessly integrating the Video & Image galleries within the text, for example, while reading the text you can watch a quick video and get right back to reading where you left off. The nine Quizzes were the next logical choice and tested the reader's knowledge on various subjects.

For the next challenge, we selected bolt-on micro apps like the Morse Code Generator which has pre-installed alarm signals but also lets you enter your own messages which are translated into Morse Code and transmitted through the flash-light function. We also added Camping Checklists and the Sun compass, which an outdoorsman would find useful and easy to use while on a hike.

The Results

"Found this app handy when my car broke down on the way home from work in the middle of nowhere. It kept me from freezing to death. So much information. More than worth the $$$ I paid for it." – Rob Taylor, Australia, Sept. 26, 2010, SAS Survival Guide app user.

It's truly inspiring when your work actually helps someone. It's probably worth more than all the accolades we have won. Building the SAS Survival Guide has been a wonderful journey. The app has been well received with over 1,000 App Store ratings averaging five stars, downloaded over 200,000 copies and has won numerous accolades from Apple, FWA etc.

The SAS Survival Guide has been selected the "iPad App of the Week", "Featured App", "Staff Favourite" and "Number One Lifestyle App" in almost all Geographic Locations.

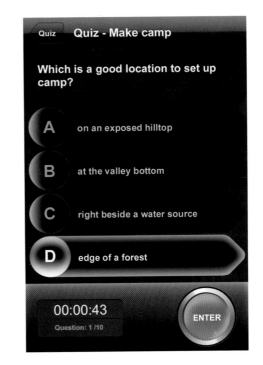

Hello Baby:
Pregnancy Calendar

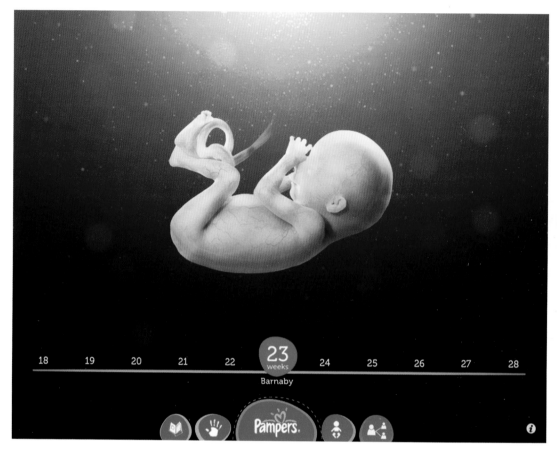

"Hello Awesome App – Wow. This is an amazing app for all moms and dads too (like me). I love that you can share the joy of parenting right from the start. The images of the baby and the hot spots interaction show off this new iPad experience. Thank you."
RanaF, App Store user review

Client
P&G – Pampers

Credits
StrawberryFrog
www.strawberryfrog.com

Awards
FWA Mobile; Cannes Cyber Lions,
Shortlist; Jay Chiat, Shortlist

The Brief
Pampers wanted to increase brand relevance with modern prenatal moms. They saw an opportunity to target highly mobile moms-to-be who are digitally active looking for content and information in preparation for motherhood and to connect to their unborn baby. A new mobile platform was launching, Apple's iPad, and Pampers wanted to take the opportunity to connect with this leading-edge, early-adopting audience to build equity as an innovative brand.

The Challenge
While Pampers is a leader in the diaper category, prenatal moms do not necessarily have a connection with the brand until the baby arrives. We knew the majority of women would be open to connecting with baby-care brands early in their pregnancy and the iPad presented a valuable relationship marketing opportunity. The challenge was to break through the clutter of the existing prenatal digital and mobile content, offering something different, desirable, and remarkable in this new mobile platform.

The Solution

We leveraged the iPad's intimate and interactive touchscreen format to build a joyful emotional experience that went beyond existing digital pregnancy content. We used technology to create something intensely emotional – Hello Baby.

Hello Baby, the only pregnancy calendar created specifically for the iPad, is a customizable, 3D-like sight and sound experience that brings to life a baby's unique development in simulated, life-size. We created an opportunity for Pampers to enable a mom-to-be to feel like she can reach out and touch the baby inside her.

Mom can personalize her baby's details and see her baby grow and develop. The multi-sensory features like heartbeat audio at the 11th week help her feel connected. She can also share her pregnancy experience with others using easy direct links to social networks.

The Results

Despite no media support and limited awareness for Hello Baby, it was the number one most popular iPad app in the Health/ Fitness section of the iTunes store for weeks. It remained in the Top Ten most popular apps in its category for months and has been downloaded and enjoyed over 50,000 times.

Pampers succeeded in opening a new dialog with consumers and demonstrated the progressiveness of the brand. In qualitative research, Hello Baby strengthened brand perceptions of Pampers as a tech-savvy, innovative thought leader and stretched the brand beyond the diapers/wipes business.

Moms are excited about this new way of experiencing pregnancy. Our iTunes app rating is four out of five stars, with incredibly positive reviews among our leading-edge target.

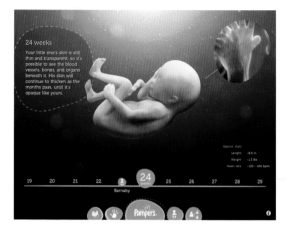

50,000 Downloads

458,912 Lines of code

150 Downloads per day

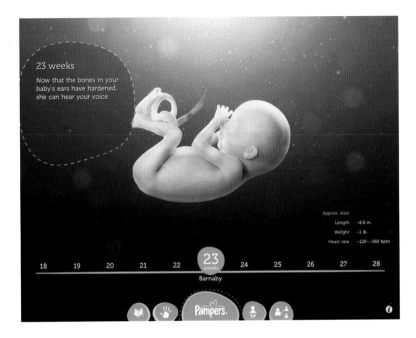

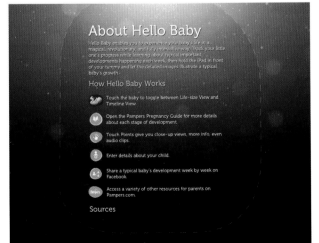

Mill Colour

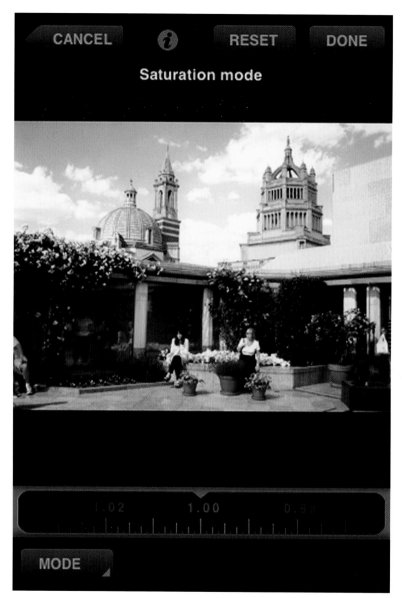

"Such a fantastic app, an essential app for anyone who's into taking photos with their phone (and making them look good!)"
m571 – App Store user review

The Brief

The brief for this project was simple: how can we give the public access to the techniques of The Mill? The best way to do this seemed to be through mobile. The iPhone's functionality, screen resolution and colour reproduction gave us a platform to bring high-end production techniques to a digital device.

Then we needed to choose a creative discipline to focus on. We decided that Colour Grading would be a really useful facility to have in your pocket – we all take photos and we all like them to look nice. But what if we could make our photos look like they were graded by pros?

So armed with this idea, we went on a mission to make the application as accessible and effective as possible.

The Challenge

Adapting the tools and terminology of high-end colour grading presented a real challenge to the team. We needed the application to be easy to use and understand. But we also wanted users to produce excellent results and learn about colour. We felt it was important to avoid dumbing down the language; this application had to feel authentic to The Mill and respect the audience. Our goal was for users to feel that they were learning the fundamentals of colour that every professional knows.

The technical challenges, beyond the usual development headaches, involved sourcing the best "looks" that would work across a multitude of exposures. Also, the team had to ensure that the algorithms employed to adjust colour were as subtle as those in a high-end suite. Quite a tall order.

Client
The Mill

Credits
The Mill
www.themill.com
Idea/Project Manager:
Andy Orrick
Design:
Ian Wharton
UE/Developer:
Nick Ludlam
Support Functions:
Jonathan Brazier
Colour Experts:
Colourists

Awards
Macworld: Best Filters App, 2010

05 Utilities

620,000 Downloads

503 Flickr members

2,152 Flickr pictures

The Solution

Unlike its competitors, Mill Colour allows the user to adjust the base image. This means that the user can achieve real subtlety in colour correction. Most filter apps allow the user to apply and adjust the filter only. But if the base image is under-exposed, no amount of filter adjustment will make the image look good.

To keep the app authentic, we used the same terminology and tools that you'd find in Colour Suites at The Mill: lift, gamma, gain, saturation and additive/subtractive colour (RGB). We wanted users to be able to make fine adjustments to their image, so we purposely stayed away from blunt tools like "contrast" which you'd never find in a professional suite.

The Results

We have shared this small piece of The Mill with over 620,000 people. The app was released in June 2009 and within a few weeks had reached 250,000 downloads.

The tool is both simple and effective, enabling all users to produce professional-looking results. The feedback was universally positive; users appreciated getting access to such high-end functionality.

To help users get the best results from their app, we included useful hints from our Colourists, as well as an archive of commercials graded at The Mill. We also set up the Mill Colour Image Gallery on Flickr which enabled users to share their work. This group now has over 500 active members.

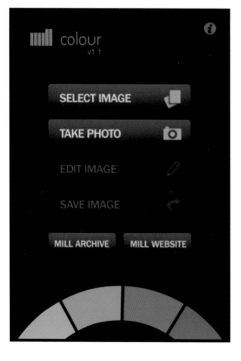

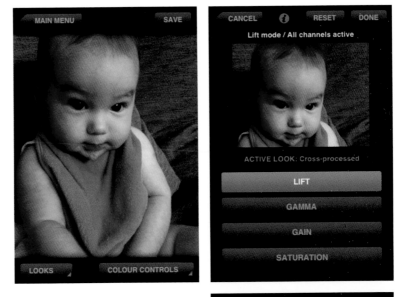

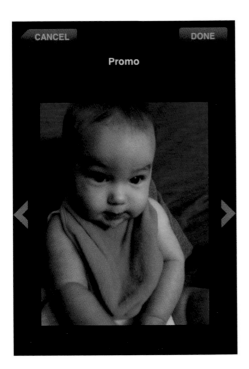

TiltShift Generator

"Simply amazing. I downloaded this on Monday
and cannot stop taking photos and editing them.
It works flawless with the high res iPhone 4 camera.
And I love how you can pick the size of the export.
This is now my number one editing app."
<u>Anthony Esposito</u>, App Store user review

The Brief

TiltShift Generator is an iPhone photo-editing application that enables users to apply lens blur and colour effect. The primary function of the app is a lens depth-of-field effect to make fake miniature and DSLR-like images. It also has simple colour adjustment, saturation, brightness and contrast control. In addition, users can post edited images to Twitter or Facebook.

The Challenge

A basic photo-editing app is huge and complex, and most such apps have lots of buttons which make it difficult to use. What we needed to do was make the application both simple and powerful. The most important thing was the user interface and deciding what we should not add to the app.

Client
Art & Mobile

Credits
Art & Mobile
www.artandmobile.com

Awards
FWA Mobile; Best App Ever Award 2009 – Best Photo Manipulation App, second place; Adorama APPOS Award 2010, the top 10 winning Apps

05 Utilities

The Solution

First we built a prototype with Flash and released it as a web and AIR app. Through online beta tests over six months, we defined the function of the app and made a simple user interface. The Flash App also produced a huge amount of buzz on Twitter before it was released. We only made a few ads and there was little pre-release, but it spread all over the world.

The Results

We developed a combination of a tab and pop-up sub-menu user interface. The five tabs are File, LensBlur, Color Adjustment, Color Setting, and Save. All the user needs to do is just move from the left tab to the right tab and use the slider for control, it's pretty simple and powerful.

370,000
Paid version
downloads

720,000
Free version
downloads

#1
Japan App Store

Opera Mini

"I love Opera Mini because there's nothing 'mini' in it. It gets me everywhere anytime I want and that's why it's a part of my everyday browsing. I don't always have to carry a laptop with me when I have Opera Mini in my pocket!"
Student, Finland

The Brief

Once upon a time, mobile web was a slow and unpleasant experience consisting of just plain text and no images. We wanted to retain as much of the quality of the web as possible and be able to deliver a PC-like mobile browsing experience to any phone. The challenge was to bring the full Internet to a simple phone with a small screen. We needed to figure out how to compress the pages, make them smaller and yet easily readable.

So, we brought in the servers that do most of the heavy-lifting for users. Engineering-wise, it's making things small that poses the real challenge.

The Challenge

One of the major challenges of course was to make the technology work, so that the compression would work on basically any mobile phone, yet keeping the full Internet experience. Another challenge we faced was convincing people to start using the web on their mobile phones. In 2005, when Opera Mini was launched, mobile Internet was known to be expensive, slow and not very user friendly. It took some time to convince people that mobile browsing can be fun, cheap and easy.

Client
Opera Software

Credits
Opera Software
www.opera.com

Awards
GetJar Gettie; Gullkorn; About.com Reader's Choice Best Mobile Browser; Quick Heal NDTV Tech Life Award; The Lifestyle and Gadget Award

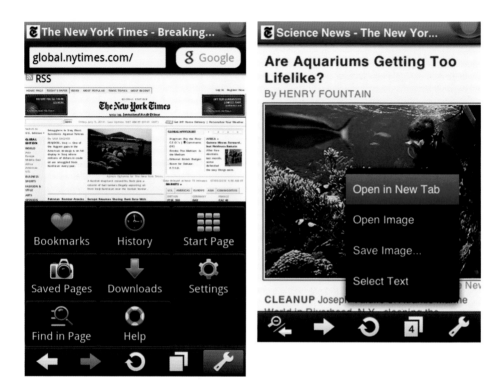

The Solution

Desktop-like browsing experience: Opera Mini offers users a more advanced browsing experience because it borrows many features and functionalities from Opera Software's browsers for smartphones and computers.

Remote processing of web data: Opera Mini shifts the heavy-lifting of downloading web-pages away from the phone (especially more basic feature phones with limited capacity) to Opera Software's servers.

Faster browsing: Opera Mini's unique server side compression technology means faster browsing, because the compressed page takes less time to load. This makes users more productive online, and relieves the frustration of slow browsing.

Lower cost: the compression reduces the size of the web-page by up to 90 percent, which means dramatically reduced phone bills for those who pay per kilobyte in data consumption.

Optimised network capacity: if each user takes up less capacity on mobile phone networks due to the compression of web data with Opera Mini, more people are able to use these mobile phone networks and at a lower cost to the mobile phone operator.

The Results

The unique solution that the Opera Mini mobile web browser provides is server side compression technology, which shifts the heavy-lifting of downloading webpages away from the phone to Opera Software's servers.

Instead of requiring the phone to process a web-page, Opera Mini uses a remote server to pre-process the page before sending it to the phone.

Opera Mini compresses web-pages by up to 90 percent (i.e. shrinks them to as little as ten percent of their original size) and reformats them using Small-Screen Rendering for easy and fast browsing on small, mobile screens.

As the server side rendering puts less strain on the mobile phone and therefore does not require a phone with large system resources, Opera Mini allows nearly any mobile phone to run a web browser – even the most basic, low-end feature phone. In many markets, computer and broadband penetration is low, while mobile phone ownership is high.

The mobile phones that many people own are basic feature phones, and not advanced smartphones. Opera Mini provides a way for people in many markets to get online, where they may not have other ways to do so.

For the end-user, this means faster browsing (because the compressed page takes less time to load) and dramatically reduced phone bills for those who pay per kilobyte in data consumption (because the compressed page in smaller in size).

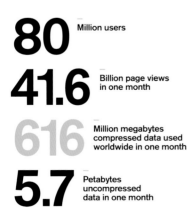

80 Million users

41.6 Billion page views in one month

616 Million megabytes compressed data used worldwide in one month

5.7 Petabytes uncompressed data in one month

HelvetiNote

Death To Marker Felt™

HelvetiNote™

"Many of the note-taking apps for the iPad feel busy or are full of distracting features, but HelvetiNote is a simple minimalist option. It lets you hide everything and just focus on your writing."
Rosa Golijan, Gizmodo

Client
Cypher13/Rage Digital

Credits
Cypher13
www.cypher13.com
Rage Digital
www.ragedigital.com

Awards
FWA Mobile; Main Page Feature
on UK and USA App Store

The Brief

With the launch of the iPad, note-taking on the device left much to be desired. Apple's Notes, set in Marker Felt, bound in faux-leather, and relying upon hand-drawn iconography was not for us. Numerous other applications were released on the platform, perhaps to improve upon Notes but most came with far too many features, complicating the simplicity and purity of note-taking. An elegant and refined note-taking application, capable of capturing a sketch or drawing, organizing thoughts, email-ready, and employing a highly legible, aesthetically pleasing typeface was in demand.

The Challenge

Given our desire to take, organize, and share notes on the iPad, we set out to design and build a note-taking application to the functional and aesthetic standards we desired. Based on our utilitarian sensibilities and early-to-market understanding of the platform we challenged ourselves to improve upon the available App Store options.

HelvetiNote™

Search

3 Notes

Welcome To HelvetiNote!

Helvetica Facts

1. Helvetica was developed in 1957 by Max Miedinger with Eduard Hoffmann at the Haas'sche Schriftgiesserei (Haas type foundry) of Münchenstein, Switzerland.

2. Originally called Neue Haas Grotesk.

3. In 1960, the typeface's name was changed by Haas' German parent company, Stempel to Helvetica (derived from Confoederatio Helvetica, the Latin name for Switzerland) in order to make it more marketable internationally.

Historically Significant Logos Set in Helvetica -

3M / American Airlines / American Apparel / AT&T / BMW / Jeep / JCPenney / Lufthansa / Microsoft / Orange / Toyota / Panasonic / Motorola / Kawasaki / Verizon Wireless

HelvetiNote™

i   ⊗   T   ✏

Welcome To HelvetiNote!

→ Helvetica Facts

1. Helvetica was developed in 1957 by Max Miedinger with Eduard Hoffmann at the Haas'sche Schriftgiesserei (Haas type foundry) of Münchenstein, Switzerland.

2. Originally called Neue Haas Grotesk.

3. In 1960, the typeface's name was changed by Haas' German parent company, Stempel to Helvetica (derived from Confoederatio Helvetica, the Latin name for Switzerland) in order to make it more marketable internationally.

Historically Significant Logos Set in Helvetica -

3M / American Airlines / American Apparel / AT&T / BMW / Jeep / JCPenney / Lufthansa / Microsoft / Orange / Toyota / Panasonic / Motorola / Kawasaki / Verizon Wireless

⚙   ✉   ⊗

1.4
Million sessions

36
Countries
downloaded

2.17
Years spent
open

The Solution
HelvetiNote – set in Helvetica. A refined set of experiences was very intentionally designed and built to take advantage of the platform's native functionality. Touch-based drawing and erasing were integrated to supplement written note-taking. Both written and drawn notes were optimized for email. A navigation scheme centered on simplicity and maximizing screen real estate was integrated. A collapsible Notes Panel was included for easy organization and a Themes Panel enables users to customize their very own HelvetiNote interface.

The Results
For a cross-studio, internally motivated project, HelvetiNote has exceeded all of our expectations. Sales have been strong and application adoption has been far and wide. Internally, both Cypher13 and Rage Digital rely extensively on HelvetiNote to meet all of their collective note-taking needs. Apple's integration of Helvetica to their native Notes application could certainly be considered a nod to HelvetiNote.

HelvetiNote™

Search

3 Notes

HelvetiNote Features
Rage Digital
Berger & Föhr

Welcome To HelvetiNote!

Helvetica Facts

1. Helvetica was developed in 1957 by Max Miedinger with Eduard Hoffmann at the Haas'sche Schriftgiesserei (Haas type foundry) of Münchenstein, Switzerland.

2. Originally called Neue Haas Grotesk.

3. In 1960, the typeface's name was changed by Haas' German parent company, Stempel to Helvetica (derived from Confoederatio Helvetica, the Latin name for Switzerland) in order to make it more marketable internationally.

Historically Significant Logos Set in Helvetica -

3M / American Airlines / American Apparel / AT&T / BMW / Jeep / JCPenney / Lufthansa / Microsoft / Orange / Toyota / Panasonic / Motorola / Kawasaki / Verizon Wireless

Aa Ee R

05 Utilites

Stickybits

"Haha I downloaded this app to win a free t-shirt
by scanning a Ben and Jerrys ice-cream.
But after that I couldn't stop scanning things
lol very fun me and my bro are competing
right now I'm winning."
Jesse Carrillo, user on App Store

The Brief
How can we let objects tell stories?
How can things talk to people?

The Challenge
We wanted to create a space where people
could communicate with each other and
share information through the everyday things
they interact with, the things that aren't
connected to the network. To accomplish that,
we needed to figure out how to attach digital
content to physical objects. The biggest
challenge was how to connect the two. RFID
allows you to attach electronic tags to
objects, but isn't widespread. What could
we create that everyone could use?

Client
Stickybits

Credits
Billy Chasen,
Founder, CEO
Seth Goldstein,
Co-Founder, Chairman

Awards
FWA Mobile

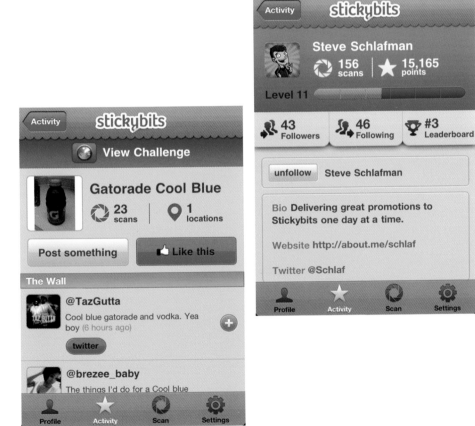

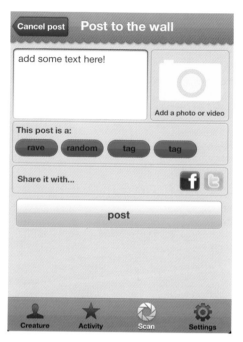

The Solution

Smartphones and bar codes turned out to be our solution. Smartphones are almost ubiquitous, and every smartphone has a camera. Stickybits uses that camera to scan the standard UPC codes found on millions of everyday objects. We currently have apps for iPhone and Android devices. We also allow anyone to generate their own code to attach to any object they want to. Once the object is scanned you can read the information left there by others – photos, videos, text and audio messages. You can also add your own message. For example, scan a book and leave a review, scan a cereal box and attach a treats recipe, or scan a wine bottle and leave a tip for the best food to pair with it. In December 2010 we added promotions to Stickybits, now allowing brands to attach deals and discounts to products. This created a solution for how we could monetize this platform. Some of our launch partners included: Ben & Jerry's, Elmer's Glue, Fiji Water, and Pepsi.

The Results

The Stickybits platform has been warmly received by our users who continue to come up with many creative ways to use it. We have people attaching bar codes to business cards that, on scan, reveal their resume. The San Diego Zoo placed bar codes at reptile exhibits and attached digital information to the codes to help visitors learn more about the animals. London Gatwick Airport printed giant bar codes on walls that, on scan, reveal videos that show the progress of the construction going on behind them. One man in Colorado even proposed to his girlfriend using Stickybits. He sent her on a scavenger hunt placing Stickybits bar code stickers all over the city, each with a message about where to go next, at the final meeting place he met her with a ring. On the brand side, business continues to grow as more brands seek out ways they can communicate with their customers through Stickybits. The promotions have been successful as users are surprised and delighted when they scan an object and win a reward.

On November 17, stickybits ran a promotion for a free $10 iTunes gift card to the first 1,000 people that scanned an Altoids tin. In a matter of hours, all cards were given away. There were thousands of scans, hundreds of posts, tweets, and shared links.

The result: when giving a highly valuable promotion to a large set of users, there is a high and instant engagement with the products involved in the promotion.

Stats over three days:
• 1,623 scans
• 1,339 unique
• 349 wall posts

Scanning locations: 32 different countries including England, Germany, Australia, Singapore, New Zealand, Israel, South Korea, and Pakistan.

1,623
Scans Altoids promo (three days)

1,339
Unique scans (three days)

349
Wall posts (three days)

32
Country locations (three days)

Kinetic: Activity Tracking & Timing

Kinetic for iPhone
Activity tracking & timing

Running

Untitled

Nov 16, 2010

Time
00:55:57

Distance
4.70 miles

Calories burned
548 kcal

Average speed
5.04 mph

Maximum speed   Average pace
6.49 mph   11.9 min/mile

Speed / Altitude / Distance

"Packaged in a beautiful interface with features to amaze and only grow with time, Kinetic is a must for every iPhone... It's a great app, you guys are on to something big!"
Filip Visnjic, CreativeApplications.net

Client
Mothership Software Ltd

Credits
Mothership Software Ltd
www.wearemothership.com

Awards
FWA Mobile

The Brief

We wanted to create an iPhone app for tracking and timing movement-based outdoor activities using the iPhone's GPS. There were already a number of apps that do just that, but they were either poorly designed or did not have the depth of functionality we wanted. Crucially, none of these apps allowed the user to choose what information they did or didn't want to display.

The Challenge

Kinetic had to be functional, expandable and customisable as well as beautiful.

The challenge was to give users the ability to display different information depending on what type of activity they are doing. For example, a runner might want data regarding time, distance and pace. And if they put the phone in their pocket then that information has to be audible rather than visual. If they are trekking, they might want a map and compass, as well as time and distance information. The user might also want this data in a different unit of measurement from the next person. We wanted to allow the user to set up Kinetic to best suit their own needs and preferences – the potential combinations were endless!

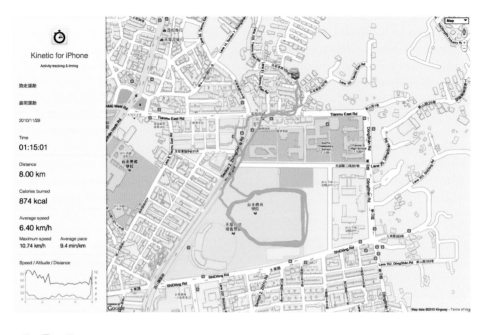

10
Months creation
time

3,000
Miles of testing

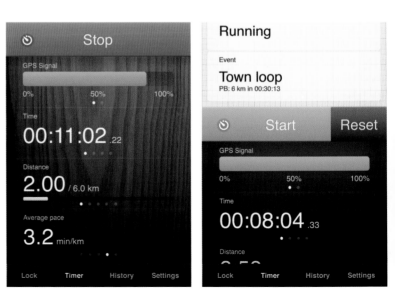

The Solution

Kinetic had to be modular, in more ways than one. We created building blocks – the modules – so users could more or less build their own individual app. Kinetic's modules provide a range of different functionality with bespoke feedback and information. There are modules for speed, distance, time etc. We also added extra functionality such as a compass, maps and graphs. The user can add the modules however they like and each module can be configured independently of the others.

This makes Kinetic highly configurable but, just as important, expandable too. To do all this we had to take a different approach to the code – we built our own framework which allows us to make additional modules available in the future through app updates as well as in-app purchases. We think our approach is quite unique.

The Results

We succeeded in creating an app that is useful, usable and beautiful, and the five star ratings, excellent feedback, support and recognition we are getting from our customers is hugely rewarding.

Kinetic's modular approach, slick design and custom UI sets it apart from other GPS tracking apps and allows us to continue to develop additional functionality and features in response to customer demand.

We can see from our analytics that Kinetic is being used on a regular basis – it's not a "one hit wonder", throw-away app, but something that our customers will keep coming back to again and again.

helloflower

"Blooming marvelous! This is a fantastic app
that delivers all it promises. It is both fun and
instructive. It turns you into a designer. Like
all addictive apps it is simple to use and the
products are quite spectacular. I use it to create
my designs and my quite young grandkids create
theirs. It gives reason to owning an iPad."
Jimdellit, from Australia

Client
HelloEnjoy

Credits
HelloEnjoy
www.helloenjoy.com

Awards
FWA Mobile

The Brief
We love creating interactive 3D graphics for the iPad and wanted to make it accessible for everybody. We envisioned our users sculpting with their fingers, focusing on the creative process and enjoying beautiful results in real time. Our idea was to build an application that was intuitive and easy to use, but capable of generating great visuals.

We chose to steer away from geometric shapes and try to emulate the beauty of those created by nature. And what's more beautiful than a flower bursting with colour and variety? Flowers have complex and intricate shapes that can only be fully captured in 3D, making them the ideal subject for our application.

The Challenge
Due to the nature of our application, the creative challenge was twofold:

On one side, it was vital that the application had a clean design and an intuitive interface, so it was easy to pick up and had a gentle learning curve. The creation of 3D graphics on a 2D screen frequently requires a clear understanding of many complex tools and how they work together.

On the other side, we also wanted the user creations to look good, with realistic organic forms and smooth colour gradients. We also needed a flexible solution to be able to recreate many different types of blossom.

drag handles to shape the petals

The Solution

We built a touch-based interface that would hide all complexity from the user. With the help of the shape tool, an intuitive 3D spline editor, the user can see the flower from any angle and accurately define each petal outline.

We also developed advanced rendering techniques and specific shaders to achieve realistic flower textures, allowing the user to select a combination of two colours per petal.

The user also has access to other tools that use sliders to adjust the number, curvature, openness and inclination of the petals.

And when the creation is finished, it can be saved in the flower gallery.

The Results

Reading the feedback from our users has made us realise the potential of this kind of creative tool, particularly for children. Content creation tools that are both accessible and produce high-quality results can help unleash a person's creativity, and that's what our users enjoy most.

Being our first app for the iPad, helloflower has enabled us to develop a solid technology base and a deep understanding of the device's capabilities, rounding out the studio skill set in the mobile area.

Internal projects of this kind allow us to experiment freely, learn new techniques and increase our expertise in interactive design.

22,500
Triangles per flower

24
Petals per flower

4
Texture stages per petal

drag handles to shape the petals

rotate

iORGEL

"Bygone retro on your iPad. A piece of bygone history before iPods, Cassette tapes and Records. Visually excellent with wooden casing and moving mechanism. Love the wind up action you do to start it. Sound is brilliant just like the real thing… To be honest it's even better than the real thing as you can add and create tunes for it."
User from UK

Client
allm Interactive

Credits
allm Interactive
www.allminteractive.com

Awards
FWA Mobile
KT 2010
Econovation App Fair

The Brief

Heavenly joy in a music box! Bring back a special moment of your life with iORGEL!

Most people in this world have childhood memories of playing with a music box. It's amusing how many of these people reminisce about the past when they hear the sound of a music box. iORGEL is a special music box that allm Interactive reinvented. Even though iORGEL is an application that is played on a digital device, it has a handmade quality that leaves warmth in people's hearts. Happiness, warmth, memory and love are some of many keywords users all over the world feel and share. Many music box fans can now create, edit and share their own songs. There are tons of new songs that are being created and shared even at this moment.

The Challenge

iORGEL is a musical instrument – therefore, the most challenging issue was to get perfect sound quality. If you carefully listen to a real physical music box, you can tell that there are many different mixed sounds. We had to perfect the sound quality for many different situations. The instrument sometimes needs to make lighter sound and sometimes needs to make deeper sound depending on the octaves. It also needs to have long lingering sound effects and most importantly make beautiful sound.

Another challenge was to develop a user-friendly interface in the editing screen. In order for users to easily create songs, we had to mix in some common interface that is used in many composing softwares while trying to keep the music box-like quality as well as giving unique user experience.

The Solution

In order to mimic the original music box sound, we utilised tons of sound synthesisers and sound modules. After creating the closest sound, we corrected the lengths of each note to mimic the natural fade-out. If the lengths of each note are too short, the note doesn't sound right when played with others and if the length of each note is too long, the app will crash due to memory issues. We perfected the sound by going through each and every one of the 48 notes in the audio editor.

As for the edit mode interface design, it would've been easier if we just had the users use a lined music sheet to create songs. However, we worked hard to mimic the original music box by creating a grid system for users to punch in the notes similar to the real physical player.

The Results

What we want to show through this app is a sense of emotion within a digital device. We worked hard to mimic the original music box sound as close as possible as well as paying great attention to visual details even the individual movements of small gears. The result of this carefully handcrafted creation made people all over the world become attached to this newly reinvented retro device.

We get great feedback even from users with no musical background showing off their musical creation. Tons of songs are uploaded and shared every day and it is great to see how different songs are being recreated and played on this music box. This unique and beautiful analog instrument that is created for a digital device brings warmness into people's hearts.

1.3 Million downloads in five months

90,000 Songs uploaded/ shared

28,000 Liked songs

54,032 Max downloads in one day

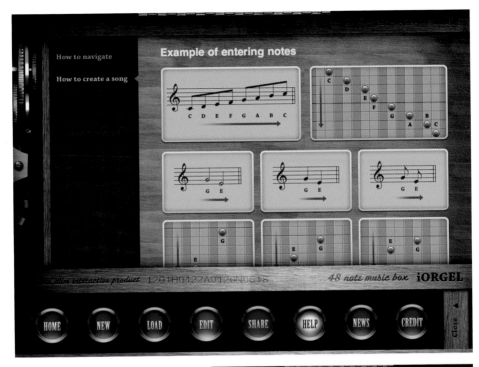

FWA Photo:
One Photo a Day

"Great to see simple ideas done well.
Awesome photos providing inspiration
through the day, no matter where you are.
Love the simple intro animation – almost
thought I had flash on my iPhone!"
Aquamedia, App Store user review

The Brief

Rob Ford from The FWA approached us to create a daily photography showcase, with the focus on one photo a day. It wasn't to be an award showcase nor was it to have any ambitions to being a reference for photography, it was simply to showcase one amazing photograph per day.

The Challenge

The challenge was to create a photo showcase which was different from those already out there. Simplicity was crucial whilst also being up to date with modern technology and people's new device requirements. We had to consider a desktop version and also a mobile version as well. With the upsurge in iPhone apps in 2009 we had to consider an iPhone app as a viable extra.

Client
The FWA

Credits
Concept/Design:
group94
www.group94.com
Programming:
Boulevart
www.boulevart.be

05 Utilities

3,000
Plus downloads, first week

50
Plus country downloads, first week

1.4
Million plus website visits

The Solution

The solution was to create a seamless project across three media: website, desktop and mobile. The website was created to show almost full-screen photographs, picked by a two-man team of editors from user submissions. With a focus on the photography, we kept the navigation as subtle as possible, with an extra fade for increased quality, brightness and contrast, to showcase each photo of the day at its best. The entire project was managed via our flash94 CMS.

We also created an Adobe AIR app which we called the "Lightboxer". This is an easy to install app so that when you add one of your favourite photos from the website, it becomes part of a screensaver on your desktop, which can randomly play your saved photos. There is an additional option with the Lightboxer to automatically show the day's Photo of the Day. Therefore, when you switch on your computer, you can see today's photo immediately.

The iPhone app makes viewing FWA Photo whilst you are on the move easy. It's even possible to add a photo from the iPhone app to your Lightboxer when you are out and about, maybe in a boring meeting, and then when arriving back at your computer, that same image is full-screen on your computer.

The Results

On the day of launch, there were 773 downloads of the iPhone app from 42 countries. In its first week, the iPhone app had over 3,000 downloads from over 50 countries. Now, two years later, it's still a very popular free photo app. Feedback has told us that many people have FWA Photo as their homepage on their browser. The website has since received over 1.4 million site visits. We have also found a complete copy of the website in another language… this tells us we must have got things right!

FatBooth

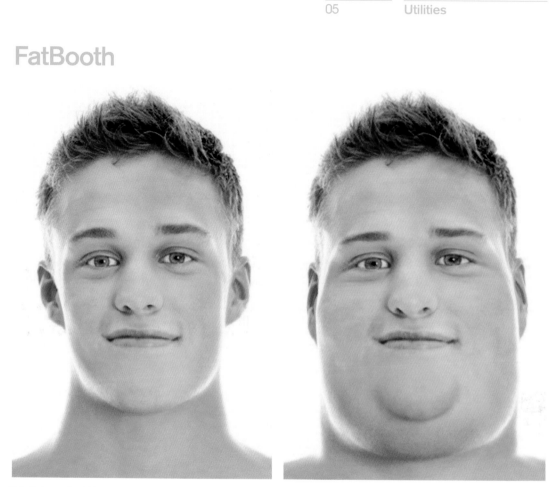

"FatBooth iPhone app re-imagines you… 100 pounds heavier."
<u>MSNBC</u>, *The Today Show*

The Brief

In early 2010, we launched AgingBooth, an entertainment app to instantly age face photos on an iPhone, with a face detection and transformation system. Following the success of this app, which topped rankings in many iTunes App Stores in the world, we were willing to find another effect at least as funny and interesting for users as AgingBooth. We thought that after having seen them old, they would like to see them fat.

The Challenge

With FatBooth, we had to allow users to instantly get their face fat without hours of make-up as they do in movies or on television to achieve this effect. Once again, simplicity of use and realism of result were our top priority. The process is very simple: you just take a picture with your iPhone, follow some easy instructions and the result is immediately shown.

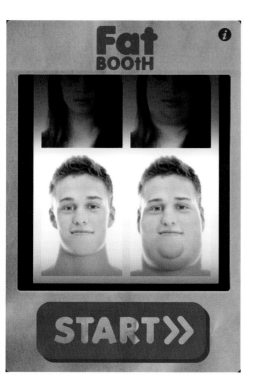

Client
PiVi & Co

Credits
PiVi & Co
www.piviandco.com
Photography:
fotolia.com
(Yuri Arcurs; CURA Photography;
Piotr Marcinski)

Awards
FWA Mobile

05 Utilities

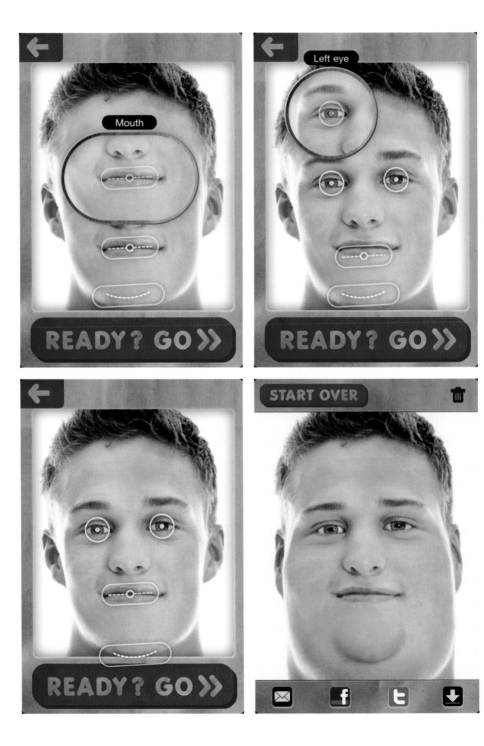

The Solution

We have perfected our face-processing system to adapt it for the different conditions in which the user has taken the original photo, so that realism is not thereby altered. Our design process was iterative, making many round-trips between technology and design to make the result as balanced as possible. Concerning graphics, we chose to evoke the world of fast food that was suitable for this app for obvious reasons, enabling the use of colour and playful forms.

The Results

Very quickly, FatBooth climbed the Top Paid rankings in many countries due to its integration of shared photos via Facebook, Twitter or email and also to the many good users' reviews on the App Store. Good reviews in major media such as MSNBC's *Today Show* in the US or the *Daily Mail* in England, and its being used by celebrities who posted their FatBoothed photo online, have also contributed to that rapid climbing in the rankings. And a few months after its launch, FatBooth is still well ranked in the worldwide Tops.

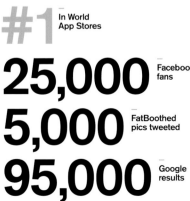

#1 In World App Stores

25,000 Facebook fans

5,000 FatBoothed pics tweeted

95,000 Google results

Afterword

by Lars Bastholm, Ogilvy

Afterword Afterword Afte
Afterword Afterword Afte
Afterword Afterword Afte
Afterword Afterword Afte
Afterword Afterword Afte
Afterword Afterword Afte
Afterword Afterword Afte
Afterword Afterword Afte
Afterword Afterword Afte
Afterword Afterword Afte
Afterword Afterword Afte
Afterword Afterword Afte
Afterword Afterword Afte
Afterword Afterword Afte
Afterword Afterword Afte
Afterword Afterword Afte

I got my first mobile phone in 1995. It was huge. It made phone calls, and that was it. Still, I was totally in awe that I had one. It was a Nokia 2110 phone with a green screen. I had to get a man bag just to carry it around, as it was way too big for a pocket. If you put it in your jacket pocket, it looked like you were packing heat. Of course that was nothing compared to the size of my next phone, the Nokia Communicator, which, incredibly for the time, allowed you to access your email. The quid pro quo was that it weighed the same as a small dog. Since then I've had more Nokias, a Sony Ericsson, a Motorola, a Palm Treo, a BlackBerry and, of course, every single iPhone ever made. The biggest change, aside from the dramatic reductions in weight and size, is that 'phone' is increasingly a misnomer. They are basically ultra-portable computers that happen to make phone calls, when you need that function.

I don't think it was completely off when the media referred to the first iPhone as the 'Jesus phone'. It may not have been the first coming, but it certainly changed everything for consumers and marketers alike. Advertising and gaming on phones used to be terrible. It was primitive to say the least. Mostly SMS-based competitions and dumbed-down versions of old arcade games. No one ever chose a phone for entertainment reasons until the iPhone arrived. As you can tell from the number of case studies in this book that are about apps on the iPhone, the device certainly inspired a new generation of creators to think about what would work for consumers on this new platform.

"Portable got bigger and more intuitive, as tests showed that both a two-year-old and a 100-year-old intuitively could figure out how to use an iPad."

A few years later, we got the iPad, and mobile marketing and communication ceased to be tied to the idea of a phone. Portable got bigger and more intuitive, as tests showed that both a two-year-old and a 100-year-old intuitively could figure out how to use an iPad. That was certainly not the case, when it came to any other tablets.

That's the beauty of many of the case studies in this book. They are incredibly simple games/ads/utilities. If anything that seems to be the secret of successful mobile executions today: it is about doing ONE thing really well. Many websites end up suffering from what I call 'functionalitis': the over-abundance of things to do and ways to engage. Stuff showed in there because you can, not because any users want it. Mobile apps are the opposite. They're lean, mean and ready to perform.

In fact, one of my favorite branded apps ever is Charmin' toilet paper's Sit or Squat, a crowd-sourced public toilet review app. If you've ever been stranded in an unfamiliar city, low on public restrooms or equipped mostly with truly scary ones, you'll appreciate the need for an app like this. And if you've ever tried to market toilet paper, you'll appreciate the marketing genius of this solution.

"As someone who lost weeks of time fighting the dastardly egg-stealing pigs in Angry Birds, I can certainly attest to the higher quality of games on the mobile platforms."

Gaming has also come a long way. While playing *Snake* isn't exactly boring, it isn't quite one of those games that take over your life. As someone who lost weeks of time fighting the dastardly egg-stealing pigs in Angry Birds, I can certainly attest to the higher quality of games on the mobile platforms. According to Wikipedia, I am not alone. With more than 50 million downloads of the game for various mobile platforms, it is, right now, December 2010, the most popular game in the world.

But even as technology has evolved and content has become much more engaging, involving and social many of the most promising attributes of mobile computing are still in their infancy.

I'm personally very interested in seeing how we will use location-based real-time communication. We're seeing the baby steps with apps like Foursquare, where you get an offer of a special nearby, when you check in. However, both Foursquare and their main competitor Gowalla seem to be struggling to find great reasons for people to check in and share their location and after a massive initial uptake amongst the digerati, it remains to be seen if either service will ever convince the mainstream to take part.

The nightmare version of location-based communication is the one we saw in Steven Spielberg's *Minority Report*, where Tom Cruise's character was being bombarded by offers, as he was walking through the mall, some based on his previous shopping history, some just pure noise. The mobile computers feel more personal than any computers ever have before, because we carry them around on our person. That makes spam feel like a personal insult, and our tolerance for it is much lower than it has been on desktop computers. On the other hand, it's easy to imagine your mobile device constantly querying your surroundings to find that pair of jeans you are looking for at the price you're willing to pay. Such user-initiated location-based scenarios will undoubtedly become a large part of our near future.

Similarly QR codes and augmented reality, as exemplified by e.g. the Stickybits and Macmillan Coffee Finder case studies in this book, are in their infancy. They could both become part of the fabric of everyday life (like QR codes have in Japan), or they could just be stepping stones on the way to something even more exciting and useful. What's so fascinating about all of this is that we can't really draw upon experience from desktop computers to learn what will work for consumers on a mobile platform. For example, I'm doubtful that Angry Birds would have been much of a hit on the Xbox, but for whiling away ten minutes, while you're waiting for the bus, it's perfect.

"With an estimated 2.8 billion 3G subscriptions by 2014, almost half of the world's population will have access to on-the-go high-speed data and the apps and other software that come with it."

With an estimated 2.8 billion 3G subscriptions by 2014, almost half of the world's population will have access to on-the-go high-speed data and the apps and other software that come with it. There has never before been a way to reach this many consumers with one piece of software or an app. So if you've been wondering why Apple, Microsoft and Google are slugging it out for dominance in the mobile field, there's your answer.

When we talk about mobile, we tend to forget that it's only been about 15 years since mobile phones became affordable for regular people. It wasn't until the Noughties that SMS/texting really took off, even if the phones have always had the ability. We've only been making video calls for a year or so. Android's only been around for a couple of years, and we've had the iPad for less than 12 months.

Mobile computing is a wide open and brand new field. There's plenty of room for everyone to play, so I only have one piece of advice for you: Go make something awesome.

Lars Bastholm
Ogilvy

Bio.
Lars Bastholm
Ogilvy

Lars has been working in the marketing industry for over 14 years. After starting up Grey Interactive in Scandinavia, he joined Framfab in Copenhagen as Creative Director. There, he worked on some of the world's most recognized brands, including Nike, LEGO, Coca-Cola and Carlsberg.

In 2004, Lars was hired to open up an AKQA office in New York, where he landed global AOR relationships with clients such as Coca-Cola, Smirnoff and Motorola.

Lars joined Ogilvy North America in the newly created role of Chief Digital Creative Officer in March 2009. In this role he is overseeing the creative aspects of digital across the region. In June 2010, Lars took on the additional role of Chief Creative Officer for Ogilvy New York, overseeing all the work coming out of the New York office. Lars is a member of Ogilvy's Worldwide Creative Council.

One of the most awarded creatives in the digital marketing industry, Lars has won a multitude of international awards, including three Cyber Lions Grands Prix.

In 2009, Lars had the honor of chairing the Cannes Cyber Lions jury and he is a frequent speaker at industry events. He was named a creative leader by the *Wall Street Journal* and has contributed to two books about the digital advertising industry published by TASCHEN. Lars is on the Ad Council's Campaign Review Committee to ensure the integrity and creativity of their public service advertising.

Lars enjoys karaoke a little too much and is a card-carrying geek who likes science fiction, bad movies, technology and extreme cooking.
–
www.ogilvy.com

"Mobile computing is a wide open and brand new field. I only have one piece of advice for you: Go make something awesome."

You would think that creating a third book in a series would be easier than the previous two but, in fact, it's a tougher process as you endeavour to make the new book even bigger and better than the previous *Guidelines for Online Success* and *The Internet Case Study Book*.

We still remember sipping tea and eating cakes again (!) in Cambridge when we sketched the concept for this third volume, just a few months after the release of *The Internet Case Study Book*. With the explosion of mobile we realised we had to give this new book immediate focus and not miss a historical opportunity to deliver a much-needed book for the entire industry. We also knew that we had to take the chance to get case studies on the new wave of mobile projects before the information was lost and new projects made the very first of these mobile apps etc. seem a thing of the past. We had to move very quickly and we did just that by fast-tracking the entire process.

We need to thank a lot of people, without whom this book wouldn't have been possible. First of all we would like to thank all contributors for their dedication and attention to the importance of the publication. You made this publication another ground-breaking one. We would also like to thank specifically Chris Allen (for his eagle eye, once again, on the manuscript), Jürgen Dubau (for the German translation), Equipo de Edición (for the French and Spanish translations), Stefan Klatte (for his work and advice on the production front) and Jon Cefai at KentLyons (for once again designing an amazing book).

A very special thanks goes to Daniel Siciliano Bretas, our right-hand at all times and the manager of the whole book. He has overseen the project at all levels from the outset and is truly a master of management and an absolute key cog in the process. His work and great attention to detail make this book another special one.

Rob Ford & Julius Wiedemann

Imprint

© 2011 TASCHEN GmbH
Hohenzollernring 53
D-50672 Köln
www.taschen.com

Design by KentLyons

Editor
Rob Ford
Julius Wiedemann
Editorial Coordination
Daniel Siciliano Bretas
Collaboration
Jutta Hendricks
Production
Stefan Klatte

English Proofreader
Chris Allen

Printed in China
ISBN 978–3–8365–2880–1

To stay informed about
upcoming TASCHEN titles,
please request our magazine
at www.taschen.com/magazine
or write to TASCHEN America,
6671 Sunset Boulevard, Suite
1508, USA-Los Angeles, CA
90028, contact-us@taschen.
com, Fax: +1-323-463.4442.
We will be happy to send you
a free copy of our magazine
which is filled with information
about all of our books.